Tetsugaku Companions to Japanese Philosophy

Volume 1

This new Springer series collects and presents studies on many facets of Japanese philosophy. Its aim is threefold: to demonstrate the unique philosophical potential of Japanese philosophy; to provide systematic and critical texts for research as well as for teaching on Japanese philosophy; and to reinforce the academic status of Japanese philosophy as an academic discipline. The series focuses on representative Japanese philosophers and on various themes in Japanese philosophy. It explores pre-modern as well as modern philosophers and themes, and provides a platform for comparisons with Western and non-Western philosophical traditions. The series reflects a growing interest in non-Western philosophical movements as well as the undeniable influence of Japanese philosophy in Asia.

This new series:

- Demonstrates the rich potential of Japanese philosophy.
- Echoes recent developments in the field.
- Enhances the academic status of Japanese philosophy.

More information about this series at http://www.springer.com/series/13638

Thomas Taro LENNERFORS • MURATA Kiyoshi
Editors

Tetsugaku Companion to Japanese Ethics and Technology

Springer

Editors
Thomas Taro LENNERFORS
Department of Engineering Sciences
Uppsala University
Uppsala, Sweden

MURATA Kiyoshi
School of Commerce and Centre for
Business Information Ethics
Meiji University
Tokyo, Japan

ISSN 2662-2181 ISSN 2662-219X (electronic)
Tetsugaku Companions to Japanese Philosophy
ISBN 978-3-319-59025-7 ISBN 978-3-319-59027-1 (eBook)
https://doi.org/10.1007/978-3-319-59027-1

Library of Congress Control Number: 2018967694

This Springer imprint is published by the registered company Springer Nature Switzerland AG.
The registered company address is: Gewerbestrasse 11, 6330 Cham, Switzerland

Acknowledgements

This project would not have materialized were it not for the energy and enthusiasm of Kevin LAM Wing Keung, who Thomas Taro LENNERFORS met during his teaching sabbatical at the University of Tokyo (funded by STINT, the Swedish Foundation for International Cooperation in Research and Higher Education). Kevin encouraged us to organize a workshop addressed at exploring Japanese ethics and technology. With the generous support of the Japan Society for the Promotion of Science (JSPS) to organize a Swedish Alumni Club Activity Seminar as well as support from the Scandinavia-Japan Sasakawa Foundation, we held a seminar at the Department of Engineering Sciences, Uppsala University, on 2–3 September 2015. The book project was a result of the seminar, and many of the presented papers are included in this book. We would also like to extend our thanks to the reviewers of the individual chapters, including Anne GERDES (University of Southern Denmark), KOGA Hiroshi 古賀広志 (Kansai University), Ana María LARA PALMA (University of Burgos), MAJIMA Takashi 間嶋崇 (Senshu University), MINOBE Hitoshi 美濃部仁 (Meiji University), and ORITO Yohko 折戸洋子 (Ehime University), as well as two anonymous reviewers of the full book manuscript for their useful suggestions and comments.

Contents

Contributors

ASAI Ryoko Department of Information Technology, Uppsala University, Uppsala, Sweden

Centre for Business Information Ethics, Meiji University, Tokyo, Japan

CHEUNG Ching-yuen Department of Japanese Studies, Chinese University of Hong Kong, Shatin, Hong Kong, China

Per FORS Department of Engineering Sciences, Uppsala University, Uppsala, Sweden

INUTSUKA Yū Graduate School of Interdisciplinary Information Studies, University of Tokyo, Bunkyo, Tokyo, Japan

Mikael LAAKSOHARJU Department of Information Technology, Uppsala University, Uppsala, Sweden

Thomas Taro LENNERFORS Department of Engineering Sciences, Uppsala University, Uppsala, Sweden

MURATA Kiyoshi School of Commerce and Centre for Business Information Ethics, Meiji University, Chiyoda, Tokyo, Japan

NAKADA Makoto Graduate School of Humanities and Social Sciences, University of Tsukuba, Tsukuba, Ibaraki, Japan

Elin PALM Department of Culture and Communication, Linköping University, Linköping, Sweden

SATO Koetsu Graduate School of Humanities and Social Sciences, University of Tsukuba, Ibaraki, Japan

SATO Maki University of Tokyo, University of Tokyo Center for Philosophy, Graduate School of Arts and Sciences, Meguro, Tokyo, Japan

SHIRAI Masato Inoue Enryo Research Center, Toyo University, Bunkyo, Tokyo, Japan

The original version of this book was revised: MS Mincho font was updated throughout the book for Japanese and Chinese characters. The correction to this book is available at https://doi.org/10.1007/978-3-319-59027-1_13

Chapter 1
Introduction: Japanese Philosophy and Ethics of Technology

Thomas Taro Lennerfors and Murata Kiyoshi

1 Ethics of Technology

Today, human life is affected, mediated, even penetrated by technologies of various kinds. Work, citizenship, leisure, physical exercise and friendship are mere examples of realms and activities which are mediated by technologies, particularly information and communication technologies. Work can be done anywhere due to portable devices and ubiquitous internet access, which is liberating but also means that there is almost no place where you can disconnect from work. Given location systems and inexpensive sensors, we can now easily track our physical exercise, both exactly how far we run, the elevation, the split times per kilometre, our pulse. With this data we can improve ourselves, but also compare ourselves to other people. Similarly, friendship has been transformed by social media, which has the unique characteristic of maintaining our weak ties, but which potentially has negatively affected our relationships to friends who are not using social media, and lead us to spend so much time on social media that our time to meet in real life has diminished.

The feeling of dazzling technological development is not new to the contemporary times. Technologies have developed gradually, but have been marked by a significant acceleration over the last two centuries. Technologies are not inevitably

The original version of this chapter was revised: MS Mincho font was updated throughout the book for Japanese and Chinese characters. The correction to this book is available at https://doi.org/10.1007/978-3-319-59027-1_13

Thomas Taro Lennerfors (✉)
Department of Engineering Sciences, Uppsala University, Uppsala, Sweden
e-mail: thomas.lennerfors@angstrom.uu.se

Murata Kiyoshi
School of Commerce and Centre for Business Information Ethics, Meiji University, Chiyoda, Tokyo, Japan
e-mail: kmurata@meiji.ac.jp

Thomas Taro Lennerfors, Murata Kiyoshi (eds.), *Tetsugaku Companion to Japanese Ethics and Technology*, Tetsugaku Companions to Japanese Philosophy 1,
https://doi.org/10.1007/978-3-319-59027-1_1

determining social changes, but any technology is usually more or less multipurpose and in many cases has beneficial as well as harmful aspects. For example, genetically modified organisms which can lead to cheaper food in larger quantities, but on the other hand results in fostering non-resilient monocultures. Nanotechnologies offer new solutions to a variety of problems also present new challenges such as privacy issues, and health and environmental issues. Artificial intelligence (AI) and robotics assist human beings for various purposes but also raise concerns both about the alleged special status and meaning of humanity and not the least the singularity – where AI liberates itself from human intervention. Nuclear power, which can be seen to produce carbon neutral energy, both presents long-term challenges regarding its waste, and the constant risk of disaster, most recently witnessed in the Fukushima disaster. Although few would question the benefits that technologies have provided, it is also necessary to reflect critically upon technological development. How does it affect the fundamental nature and living conditions of human beings? How does it affect our way of relating to others? How does it affect dreams and visions about the future?

Discussions about good, bad, and ugly technology take place in everyday conversations and in contemporary media. In the scholarly domain, the nature and impact of technology and technological development is discussed by philosophers, ethicists, sociologists, and engineers to name a few. Often, such discussions go under the labels of philosophy of technology, ethics of technology, ethics of product development, or ethics of engineering.

A first start for discussions about the ethical aspects of technology is to argue that technology is not value free. Still, it is a rather common understanding in society that technology is a mere means to an end, an instrument to fulfil a particular purpose that exists "outside" the technological artefact or system. If that would be the case, there would be no point in discussing the ethics of technology, since ethics and values would only be located in the actors using the particular technology. Underlying this idea is a quite strict separation between humans and artefacts, between the subjective and the objective, between purposeful spirit and dead matter.

However, technology does embody values. A very telling example is Langdon WINNER's discussion about the overpasses leading from New York to Long Island, designed by Robert MOSES in the 1930s (Winner 1980), although its historical accuracy has been contested (Joerges 1999). Pavements and traffic islands represent a value that cars should not go there. They indeed could drive on the pavement and traffic islands but that would lead to a risk of damage to the cars (Latour 1992). Everywhere in society there are material artefacts shaping our behaviour.

There are various kinds of technologies and various ways in which humans relate to these. Don IHDE, an American philosopher of science and technology, claims that there are four ways for us to relate to technological artefacts. In *embodiment* relations, the technology withdraws or becomes transparent when used, for example a blind man's stick, or a hammer. In *hermeneutic* relations, we read the world by means of technology, for example a thermometer. In *alterity* relations, we experience technology as something other, for example our dealing with an ATM or a robot. In *background* relations, technology like lighting, heating or cooling provides us with the *milieu* in which we are active (Ihde 1979). Similarly, Dutch philosopher of technology Peter-Paul VERBEEK argues that technologies should be seen as mediating our experience and that technology is always co-constructed with humans, in other words, one ought to take technologically constructed human beings as a starting point. Since humans are always mediated by technology, we cannot create a strict divide between subject and object (Verbeek 2016).

Such examples lead to ideas about the fact that we are indeed shaping society by designing technology. New technologies lead to new forms of human life and new interactions. This has created an extensive debate amongst philosophers and engineers about how to go about developing technology that is ethical. The first point that such discussions usually make is to care at all about ethical issues. Because the development of technology is usually undertaken with good intention, much technological development is in practice rather blind to the issue of ethics. To counteract these tendencies various frameworks have been developed to effectively integrate ethics into decision making. Examples of these are the ethical technology assessment, value sensitive design, various forms of cost-benefit analysis, multi-criteria analysis, thresholds, and so on (van de Poel and Royakkers 2011). All these frameworks communicate the need for a reflexive development of technology, which truly sees technology as creating society. When building in ethics into technology, there is also an ongoing discussion about whether to design technology to enhance autonomous, rational, human decision-making or whether to nudge, influence, persuade or even force humans into certain patterns of behaviour. This discussion should not be seen as dichotomous but rather represent different values built into technology.

Related to discussions about the design of technology, there is inevitably discussions about the use of technology. As we know, a hammer could be used to hit a nail, kill a person, or even as an object of art, as Don IHDE has it (Ihde 1979). Therefore, one should be careful to attribute full agency to the designers of technology. Often, users of technology can redefine it, enlarge or change its area of application, and modify it in various ways. The values of technology thus become redefined.

In terms of the issue of how we design, use, and even discard technologies, there is a more philosophically fundamental discussion about the way technologies influence us. Technology creates society and defines the way humans perceive the world and act in it. It is not a full determination, but one cannot neglect its impact. For example, it has been argued that our friendships on social media are less shallow, but on the other hand broader, than what used to be the case, and that rather than asking, relating, and listening to others, social media promotes observation of friends, almost similar to observing works of art in a museum (Andersson and Lennerfors 2011). Martin BUBER, the author of *I and Thou*, would probably state that there has been a shift from I-Thou relations to I-it relations (Buber 1996). This is sometimes not seen as ethics of technology, but rather philosophy of technology. However, whatever the wording of the issue, to study the impact of technology on humanity is central for ethics.

Since technology influences the way we see the world, many philosophers of technology argue that technology has had a profound impact on the history of philosophy. For example, Plato in *Timaeus* depicted the world as being the creation of a divine craftsman. One could even argue that the Aristotelian teleological structure of the world – that everything has a purpose – was devised as a result of the existence of tools. Every tool has a purpose, so even the natural and human world was seen in these terms.

That is, technology goes much deeper than the very machines and artefacts that we encounter in everyday life. A paradigmatic example of the depth of technology is evidenced in Martin HEIDEGGER's work. Heidegger (1977) took the view that technology is a means to an end, and turned it around. He argued that if we see technology as a means to an end, this very belief will condition us to try to master technology. The will to mastery becomes all the more urgent the more technology threatens to slip from human control, Heidegger puts it. Although he argues that the view that technology is a means is correct, he says that it does not capture the essence of technology. Technology, rather, is in essence about bringing-forth, about revealing. Here Heidegger sees technology in the light of an underlying way of thought. Revealing seems extremely opaque but is indeed understandable. There is even a saying that if all you have is a hammer, then everything starts to look like a nail. Another example concerns the Rhine river. Modern energy technology revealed the Rhine river as a depot of energy, rather than just a river. These are examples of how technology reveals the world. Heidegger also argues that modern technology reveals the world as a standing reserve. Rather than a river being a natural object, it now becomes a standing reserve which can be used by humans if needed. The essence of modern technology is thus not anything technological but an underlying mindset which challenges-forth man to reveal the real as a standing-reserve.

Similarly, French philosopher Jacques ELLUL saw the essence of technology in a worldview, a particular form of thrownness. Ellul uses the concept of technique, which he sees as the totality of methods, rationally arrived at and having absolute efficiency (for a given stage of development) in every field of human activity (Ellul 1964). Technique cannot be reduced to a machine or a collection of machines, but has become the dominant factor in the "technicist society" of the Western world. Ellul argues that technique now constitutes a fabric of its own, replacing nature, a *milieu* in which human beings must live, and in relation to which they must define themselves. He holds that it is a universal mediator, producing a generalised mediation, totalising and aspiring to totality (Ellul 1983). Technology is thus just an expression and by-product of technique, on proceduralising, on effectivising, on aiming for highest productivity. Technique is an ensemble, where each element only has significance within the ensemble. Technique obeys the rationality of self-augmentation, automatisation, limitlessness, and acceleration. An ethical relation to this system, according to Ellul, is not to strive to function practically within the system but to offer a global alternative to it.

Given the preceding discussion, our view of technology is a very inclusive one, which is aimed at encompassing various debates surrounding the concept of technology. Technology is both artefacts and techniques – skills, procedures, and so on. Technology is also the underlying way of thinking, explained by Heidegger and Ellul as enframing and technique. The view proposed in this book is therefore that the intersection between ethics and technology spans from the theoretical to the practical, from discussions about the impact of technological artefacts and technological thinking upon ourselves to concerns about how not only to ethically design, use, re-use, and discard technology, but also to ethically relate to and engage with the predominant technological modes of thought in today's society.

2 Ethics of Technology, Inspired by Japanese Philosophy and Ethics

By writing this book we take issue with a Western bias in studies of ethics and technology. Often scholars use approaches stemming from Western philosophy, whether it is utilitarian and deontological approaches, virtue ethics, ethics of care, social contract theories, theories of justice, etc. This also goes for the more normative views where a design choice is framed as trade-off between maximising pleasure, utility or preference satisfaction, eschew rights violations, and so on. Of course, there are several exceptions to this overarching diagnosis, for example Confucian studies of social media (Wong 2013). Still, the main ways to be theoretically inspired in the answer to ethical questions is through Western philosophy.

In contrast, this book is an attempt to read Japanese philosophy and Japanese ethics to explore the intersection between ethics and technology.[1] The editors of the volume and many of its contributors are either based in Japan or has some connection to Japan, which is one rationale for the creation of this book. However, we also believe that the Japanese tradition of philosophy has something unique to offer to discussions about ethics and technology. For example, WATSUJI Tetsurō 和辻哲郎 (1889–1960), who is admired as one of the greatest Japanese philosophers and ethicists ranked with NISHIDA Kitarō 西田幾多郎 (1870–1945), found out that the Japanese term *ningen*, corresponding to the English concept of human beings, is used to simultaneously express the world or *seken* a person lives in and a person who lives in a *seken*, and maintained that the principal problem of ethics is the fundamental dualistic structure of *ningen*, who is self plus certain relationships with others or *aidagara*, and the inseparability of human existence from *aidagara* should be understood (Watsuji 2007). Watsuji's view of ethics would enable an approach to technology from an angle different from Western views (see Chaps. 3, 4 and 8).

Japanese culture had been shaped under the influence of Chinese and Indian ones since ancient times. More precisely, Japanese philosophical, ethical and aesthetic thought had been developed through the process of arrival/import of thought from overseas, mainly from China (even Indian thought such as Buddhism); selective acceptance of foreign thought; and deepening, or often Japanisation, of the accepted thought through relativising it, interpreting it and, sometimes, fusing it with existing thought. The geopolitical conditions of Japanese islands, which allowed Japan to have access to advanced civilisation without facing the direct threat of a military

---

[1] This book adopts the Hepburn system (*Hebon-shiki rōmaji*) for the romanisation of Japanese, and *hanyu pinyin* for the romanisation of Chinese. The names of Japanese and Chinese people including the contributors to this book are written putting their surnames first.

invasion, enabled such processes at least in pre-modern times (Watsuji 2011a: 26, 36; Shimizu 2014: 15–18).

Western science and technology including astronomy, geography, geometry, physics, medicine and military science started to be learnt in the early eighteenth century from Dutch books; Holland was the only Western nation authorised to trade with Japan by the Tokugawa government. At the end of the Tokugawa shogunate when the government adopted the policies to open the country, English, French and German books on science and technology were also read. However, Japanese scholars of Western studies at the time had the dualistic idea that it was important to adopt prominent Western science and technology while upholding Japanese or Eastern moral standards which were deemed superior to Western ones (Shimizu 2014: 296–297).

This idea was compelled to be changed when Japan's modernisation and industrialisation started at the time of the Meiji Restoration starting in 1867. There was a social recognition of the importance to understand the culture and social systems that supported Western civilisation. Many scholars and thinkers such as FUKUZAWA Yukichi 福澤諭吉 (1835–1901), one of the greatest enlightenment thinkers and educators in Japanese history, strained to understand and spread Western thought and culture (Fukuzawa, 2012; Shimizu 2014: 297–308). Consequently, Western culture was gradually absorbed in Japan. However, the teachings of Christianity were not spread among the Japanese although the ban of the religion enacted by the Tokugawa government in 1613 was lifted in 1873. Actually, the Christian ideas that God is the creator and that God created man in his own image have not exerted any influence over Japanese social norms (as to religious circumstances in Japan, see Chap. 7).

Recent discussions of Japanese philosophy claim that Japanese philosophy is transcultural (Lam 2011). Indeed, ABE Masao 阿部正雄 claims that in its essence it is a kind of synthesis of a "fundamental criticism of Western thought and radical reflection on Eastern thought" (Abe 1990: ix). The reason is that philosophy is an imported term into the Japanese context in the ushering in of the Meiji Era, and Japanese thinkers grappled with how to relate this term and its corresponding practice to the tradition of thought of Japan. Whereas NISHI Amane 西周 (1829–1897), an enlightenment thinker, coined the Japanese term "tetsugaku" for philosophy in 1874, NAKAE Chōmin 中江兆民 (1847–1901), a torchbearer of the Japanese Freedom and People's Rights Movement in the 1880s, even famously claimed that there has been no philosophy in Japan in his book *One-and-a-half Years* published in 1901. He insisted that Japanese studies (*kokugaku*) scholars MOTOORI Norinaga 本居宣長 and HIRATA Atsutane 平田篤胤 during the Tokugawa Period were archaeologists, ITŌ Jinsai 伊藤仁斎 and OGYŪ Sorai 荻生徂徠, who are considered most influential Confucian philosophers in the Tokugawa Era, were classics scholars, and KATŌ Hiroyuki 加藤弘之 and INOUE Tetsujirō 井上哲次郎 who were contemporaries of Chōmin's were just introducers of Western philosophy they learnt though they claimed to be philosophers. Abe (1990) presents a reinterpretation of Chōmin's claim arguing that it depends on how you define it: in the case of philosophy as a rational pursuit, then maybe it has not existed in Japan. In the case one means a philosophy of life, similar to that of Bergson, Kierkegaard, Nietzsche, which takes everyday experience as a point of departure, combining elements of religion and rational inquiry, then yes, Japan has had philosophy (Abe 1990: viii).

The meeting between Japanese thinkers and a Western philosophical tradition led to a reshaping of the Japanese tradition of thought, one that embodied an engagement with Western thinkers in its very core. Japanese philosophers at the time therefore read European philosophers' books hungrily and sought to stay in European countries, especially Germany, to study with Western philosophers, such as Heidegger in the case of TANABE Hajime 田辺元 (1885–1962), KUKI Shūzō 九鬼周造 (1888–1941) and MIKI Kiyoshi 三木清 (1897–1945), yet without disregarding the Japanese theoretical tradition, most often drawing on Buddhism and Confucianism. In fact, Nishida developed his theories through critical considerations of Western philosophy based on his experiences of Zen Buddhist practices. Kuki wrote most of his important work *The Structure of Iki* during his stay in Germany and France, in which *iki* was considered, by Kuki, as traditional aesthetics and values unique to Japanese race. Watsuji wrote the book *Climate and Culture: A Philosophical Study* inspired by Heidegger's *Being and Time* during his stay in Germany though he hesitated to continue to stay in the country.

This encounter led to William LAFLEUR (1985: xiii) stating that the Kyoto School, a loose congregation of philosophers, based in Kyoto and all inspired by Nishida, was truly comparative. This was not only due to a contingency, but the members of the Kyoto School claimed that philosophy ought to be comparative. They never desired to become obedient pupils of Western philosophers. Instead, they had a strong will to develop their own philosophy in dialogue with Western philosophical thought. There are definitely differences between Japanese and Western traditions that can possibly not be bridged by some single overarching principle. However, a creative synthesis is possible. This is echoed in SHIMOMURA Toratarō 下村寅太郎's statement of the unity of Eastern and Western thought: "The rolling of all things into one" (Abe 1990: ix). This is also the meaning of Lam (2011)'s statement that Japanese philosophy is in its essence transcultural.

While Japanese philosophy is interested in Western philosophy, LaFleur holds that Western professional philosophy has become an insular phenomenon (La Fleur 1985: xiv). He claims that during the mid-20th century philosophy became a new scholasticism feeding only on itself, but being truly ignorant about things outside the Western tradition. However, during the same decades, such boundary spanning was taking place in religious departments. Still, most of the interest in Japanese philosophy in the West does not come from traditional philosophy departments. It could be the case that the Western nature of discussions about philosophy and ethics in relation to technology derives from this alleged insular tendency in professional philosophy.

Given this transcultural nature of Japanese philosophy, one wonders what the essence of Japanese philosophy is. The same question has been asked by Lam and also by Calichman. Calichman (2005) claims that given social constructivist views on identity, whether one identifies as Japanese or indeed doing ethics is a contingent construction. One should therefore not look for any underlying substance regarding what Japanese ethics mean. However, although one cannot exactly pinpoint the meaning of Japanese ethics, it is not difficult to delimit it from other fields of thought. Watsuji (2011a, b, c, 2012) and Satō (2012, 2014) trace the history of Japanese ethical thought manifested or implied in historical records, legal codes, myths, religious books, narratives, poems, plays and so on since the early eighth century. At the present day, Japanese ethics could be seen as a subcategory of Japanese philosophy, which is in itself a constructed category containing a number of philosophers who either by themselves, or by someone else, have been classified to belong to the category of Japanese philosophy. This book draws on those thinkers including Nishida, Watsuji, MINAKATA Kumagusu 南方熊楠 (1867–1941), IMAMICHI Tomonobu 今道友信 (1922–2012), YUASA Yasuo 湯浅泰雄 (1925–2005) and ICHIKAWA Hiroshi 市川浩 (1931–2002) when analysing contemporary technological issues. But, similarly to the distinction between discourse and practice, this book also is inspired by Japanese ethical practices and how they relate to technology. The discourse would thus correspond to published works by Japanese philosophers about the nature of ethics, while practice would mean how ethics is and has been practiced in Japan.

This book is born as a transcultural encounter between researchers in the field of Japanese philosophy, connected to Kevin LAM Wing-keung 林永強 and a network of Swedish scholars interested in the ethics of technology, all of which are related to the Centre for Business Information Ethics at Meiji University. A workshop was held at Uppsala University in September 2015, where scholars of Japanese philosophy were encouraged to reflect upon the ethics of technology, and Swedish scholars were encouraged to read Japanese philosophy and let themselves be inspired by it to analyse their principal phenomena of interest.

In this encounter, several aspects of the transcultural nature of Japanese philosophy became clear. First of all, scholars who are not knowledgeable of the Japanese language can still appreciate and learn from Japanese philosophy. The Japanese language should therefore not be seen as an entry ticket to engaging with Japanese thought. As in the case of reading other works not written in English, there is of course an advantage to know the Japanese language. Although the Japanese language might not be an entry requirement, quite soon one notices that it is absolutely vital, possibly not because of the difficulties in translating from Japanese to English, but because of the mere fact that much Japanese philosophy is still not translated into English. There are a large number of commentaries and secondary literature about Japanese philosophy, but many of the original works are still waiting to be translated.

3 Social Acceptance of Western Technology in Japan

In the first part of our introduction, we argued that we can never fully separate the subjective from the material. This idea, that philosophy is always connected to material developments, is echoed in the present and the following sections where we briefly review the social acceptance of Western technology in Japan and the views of Japanese philosophers on ethics and technology.

That philosophy develops in relation to material developments does not only relate to the existence of particular forms of technology, but also to the concept of technology itself. At several times in history, Japan was exposed to technology from the West, and technological development was also resisted. In the well-known book for students of technology "Technology Matters," it is explained how Japanese warriors adopted matchlock guns introduced by Portuguese traders in 1543. Japanese swordsmiths learnt how to make the guns and improved them, and guns gradually diffused among Japanese feudal lords. In 1575, guns were allegedly decisive in the Battle of Nagashino. However, after the end of the Age of Provincial Wars in 1615 a gun became less important to the Samurai for what cannot be explained with other reasons than cultural. It was both a question of government restriction, but also due to that the dominant Samurai class rejected the weapon whereas a sword was recognised as a thing of utmost importance to them. Guns reentered only after the arrival of Commodore Perry's black ships (Nye 2006: 17).

Indeed, the most salient encounter was possibly during the Meiji era, where the connections and interchanges between Japan and the West intensified dramatically. Western technology was from the beginning considered to be vastly superior to Japanese technology and this feeling of inferiority led to interesting consequences. However, culturally and ethically, Japan deemed itself at least on a par with the West. Therefore, Japan sought to introduce Western technologies into Japan, but while keeping the spirit of Japan. Still, it was noted that the Western technologies were not just means to an end, but something truly Western in itself. Tetsuro Najita explains that technology belonged to the Western Other (Najita 1989: 11). He connects this with the previous interpretation of Confucianism into Japan, which still bore a foreign image:

> Japanese self-consciousness expressed itself with a primary reference to continuous "culture" and not to technological "work" – the latter, in the final analysis, being like Confucian knowledge attributable to the Other. (Najita 1989: 9)

The Japanese culture and Western technology were therefore posited as almost opposing forces. From such a viewpoint, an ethics of technology would even have been an oxymoron: ethics and culture were separated from technology. As the only Asian country having adopted industrial technology, the identity needed to be asserted not as a Westernised Asian country, but as a modernised Japan. Japanese thinkers in the early 1910s said that technology placed culture under constant siege (Najita 1989: 10). For example, the celebrated writer Natsume Sōseki 夏目漱石 (1867–1916) wrote as early as 1911, that the cost of rapid technological growth was enormous and painful. Within the modern Japanese bosom resided for Sōseki a crippled personality, akin to the heroine in one of Maupassant's novellas who leaps from the window to show her fidelity and is utterly dependent on the care of others for the rest of her life (Najita 1989: 11). Similarly, Tanizaki Jun'ichirō 谷崎潤一郎 (1886–1965) wrote in 1933 that technology, positivism and progress drastically reduced the ethical and aesthetic sensibilities of the Japanese (Najita 1989: 12). For these authors, culture could be seen as a form or resistance against the technologisation of society. Such tendencies still exist. For example, in the foreword to *The Nothingness Beyond God*, Robert CARTER's introduction to Nishida's thought, Thomas KASULIS writes that:

> Many issues that Nishida faced in Japan in his time have become, in a different way, issues for us in our time. We are finally feeling the culturally shattering implications of bifurcating fact and value. We live in the age of nuclear armament, genetic engineering, ecological destruction, medical technology, and even computer stock-portfolio management. Technology has vastly enlarged the horizons of what is, or at least, what could be. Yet, we are increasingly sensitive that some of what is ought not to be, some of what could be should not be. (Kasulis 1997 in Carter 1997: xvii)

Although this is one tendency that is still part of discussions about ethics of technology, Najita argues that, in the Japanese context, this problematic relation came to change after the Second World War. Rather than seeing culture under threat from technology, technological excellence came to be seen as a consequence of cultural exceptionalism, rather than the opposite. Technology was integrated into the new Japanese culture, and by and through its technology, a Japanese identity was formed (Najita 1989: 14). Japan became the leader of various technological fields, and there was a push to transform society by means of technology. Even toilets, that are often analogue and mechanical in the Western context, became an area to electrify.

4 Japanese Philosophers' Views of Ethics and Technology

Following the very brief overview of the relationship between Western technology and Japanese culture from the Meiji era, let us now turn our eyes to how several representative Japanese philosophers conceived of ethics and technology. The particular philosophies of technology mentioned in this section are not necessarily located within the contemporary context, but should be read as related to their particular time. For example, while MIKI Kiyoshi and TOSAKA Jun 戸坂潤 (1900–1945) did not see the nuclear holocaust at the end of the Second World War, NISHITANI Keiji 西谷啓治 (1900–1990) who witnessed this event had a particular interest in nuclear technologies. What happens in the world without doubt affects how philosophers theorise technology. Let us start this survey by turning to NISHIDA Kitarō.

Nishida held that technique is an expression of people's spirit as it interacts with the environment and through the interaction forms itself. He argues that we create things through technique and in creating them we create ourselves (Feenberg 2010: 120). Nishida saw our world as technological since it is a world of *poiesis*, a self-formative act that moves from the created to the creating (Murata 2003: 233; Murata 2008: 4–6). Nishida wrote that "Conceiving something and grasping it through acting intuition, means seeing it through making it, comprehending it through poiesis" (Nishida 1988, Vol 9: 194). Poiesis could here be conceived not only as technological activities but also as those of the artist. Still, a core issue is that individuals have a creative relationship to the world. Although we inherit a social body, "endowed with language and tools" (Nishida 1988, Vol 14: 290), culturally transmitted from generation to generation, there is a way to change it through creative acts.

Nishida also emphasises the radical potential of technology, not its instrumental role. "Even in the simple process of building a house, things are not given only as material but as something which has a fateful significance for our action. In every action we stand on the brink of crisis in some way or other. Our world of everyday life is a world of true crisis" (Nishida 1949: 70, in Murata 2003: 234). Crisis is here not seen as something negative but as a *Kehre*, a turning point. The poietic philosophy of Nishida facilitated by art and technology gives a quite promising view of our ways to influence the future.

WATSUJI Tetsurō was a prolific Japanese philosopher who similarly to Nishida was interested in situating Japanese thought on a global philosophical scene. He was an eager student of Western philosophy writing treatises on Kierkegaard and Nietzsche to name a few. He is most well-known (at least in the West) for his studies on ethics. He opposed individualism, which he saw as prevailing in the West, and by revisiting the meaning of human being – *ningen* in Japanese – created an ethics based on betweenness or *aida/aidagara*. As will be discussed in Chaps. 3 and 4, Watsuji, although he seldom approached technology explicitly, saw technology as a part of the *milieu* in which human beings are active. It is argued in these chapters that Watsuji saw technology as representations of betweenness. The meaning of betweenness in the field of robotics is discussed in Chap. 8.

MIKI Kiyoshi, greatly influenced by Nishida, explains society as the product of the form-giving power of imagination, and attributes a central role to technology in this process. Technology is held to be the expression of imagination in this world. Technology is subjective-objective in that it both requires human reason and creativity, but also is manifested in a concrete, seemingly unalterable form in the world. Miki holds that rather than technology just inventing new means, through technology new ends are created. In this way, history progresses. History is nothing but this movement through technological creation. Miki holds that:

> As formative action, our actions are historical. Historical actions are technological. Indeed, history is created technologically; historicity cannot be conceived apart from technology. (Miki 1967, vol. 7: 211, in Feenberg 2010: 120)

Corresponding to our previous discussion about Western technology, Miki argues that Japanese culture must reshape Western technology to conform to the Japanese spirit. In other words, to combine the spiritual elements of the East with the technology of the West.

TOSAKA Jun, also a Kyoto School philosopher, discussed the nature of technology. He saw technology and labour as inseparable, and aimed to understand technology in its context: that the human and the technological are always intertwined in larger systems. A single machine cannot operate alone (Kimoto 2009: 127). AIKAWA Haruki 相川春喜 questioned Tosaka's view by deducing the concept back to Marx's *Capital*, identifying technology as objectivist, and mechanistic. What separates Tosaka from Aikawa is his view of technology as a dynamic, mutual transaction of subjectivity and objectivity (Kimoto 2009: 131). Tosaka is discussing ideational technique as well as material technology – which very much prefigures current discussions about immaterial labour. Such subjective techniques signify the "ability in general to transform" (Kimoto 2009: 136).

NISHITANI Keiji discussed technology and mechanisation of human nature in relation to autonomy and freedom. While science and technology are developed with the hope of providing more freedom and power over nature, man rather shows a tendency to forfeit his nature and mechanise it (Nishitani 1982: 84; Davis 2011: 90). Nishitani therefore has a more pessimistic view about the potential of technology for creativity. Nishitani understands the modern crisis of nihilism in terms of a failed assertion of human autonomy that paradoxically succumbs to the dual heteronomies of exterior technological mechanisation and the interior infinite drive of self-will (Davis 2011: 90). Behind the scientific rationality and technological will of modern human being lurks the same infinite drive that the ancient Buddhist doctrine of karma sought to expose (Nishitani 1986–1995, Vol. 11: 168). Technology implies an intellectual apprehension of these natural laws. When prehistoric man learnt to make tools and to use them – for example, in making fire – this skill contained in embryo an understanding of the laws of nature *qua* laws (Nishitani 1982: 81). As the natural laws come to be understood, they are lived out through instrumental skill. They become manifest as laws through the technology of man. In the field of mechanical technology, such laws become even clearer.

> It is precisely on this field of mechanical technology, where knowledge and purposive activity make the greatest advances and work in closest unity, that the fog lifts from the laws of nature to lay bare their character as laws most clearly. (Nishitani 1982: 81–82)

In machines, the laws of nature function in their purest form, and at the same time machines manifest the human being's supreme emancipation from the laws of nature, and the supreme apparition of freedom when using those laws (Nishitani 1982: 84). However, at present, the relationship is now one of inversion. The rules of law have become present deep in man, and even human life and work as a whole has been progressively mechanised and impersonalised (Nishitani 1982: 85). In the emergence of mechanical technology, the naturalness of nature and the humanness of humans are broken down: the laws of nature come to reassume control over man who controls the laws of nature. This is a mechanisation of man, towards the loss of the human. At the same time, a human tendency has appeared that stands aside from all natural laws, a human which adapts itself to the naked vitality of life while standing on the grounds of nihility (Nishitani 1982: 86), surrounding oneself to shallow amusement or solitary withdrawal. Human life is mechanised both socially and psychologically (Nishitani 1982: 88), and the only solution is to flee into a life bound by desire.

Stenson (1989) compares Nishitani's view of technology with Heidegger's. Neither of them wishes to do away with the benefits of technology, but also see an aspect of it which works in us unnoticed and endangers the very essence of the human (Stenson 1989: 124). Science and technology transforms the world into a standing reserve of objects to be manipulated. Technology makes humanity an alter-ego of his own gadgetry. Summing up, the emergence of the machine has led to a twin tendency of mechanisation of human life and a life bound by desire. There is an imbalance in the progress of science and progress of morality (Nishitani 1982: 87), which could be counteracted by ethics, art, and philosophy (Nishitani 1982: 89).

Nishitani lived to witness the atomic bombs, and discussed nuclear technology. He argued that the state of Hiroshima right after the drop of the atomic bomb could be regarded as one end of scientific actuality in the natural world that appeared as actuality in the human world. This points to the existence of parallel actualities that are potential simultaneously. Katsuya (2014: 68) summarises: "The important thing is, however, not to select either world as true, but to search for a viewpoint from which both of them can be seen as true." Indeed, as Paul VIRILIO has argued, technology cannot be separated from its accident (Virilio 2007).

In his book *The Philosophy of Science History* published in 1941, SHIMOMURA Toratarō (1902–1995) examined the characteristics and nature of modern science based on his study on the Western academic frameworks from an intellectual historical perspective. Up to the second half of nineteenth century, he said, there had been no definite line of demarcation between natural science and philosophy; all fields of modern science were subjected to philosophy as the love of wisdom. The establishment of natural science was enabled by the development of the concept of nature in the field of philosophy and the establishment of mathematics as a separate discipline which was formed in the tradition of Western science and philosophy originated from Greek philosophy, and simultaneously that establishment of natural science as physics drove the establishment of philosophy in the modern sense that transcended the limits of natural science as metaphysics. The Western academic framework is characterised by the triune structure of mathematics, natural science and philosophy (Shimomura 1975: 74; 84–85; 217–218).

According to Shimomura, the establishment of natural science separately from philosophy promoted the development of modern technology. The role of technology as a fruit of scientific progress is to link natural science with human interests whereas scientific research tends to eschew practical purposes. Modern technology is not merely the copy of nature or the application of natural laws. Instead, it is a kind of alchemy or magic that brings about re-formation of nature through creating artificial objects such as machines and processes (Shimomura 1975: 238–241; 246). Machines are natural, supernatural and artificial, and are realised based on the modern natural view and metaphysics. Hence, the phenomenon of mechanisation is not merely a technological issue but also a philosophical and cultural issue related to societal ideals (Shimomura 1975: 254–255).

Following these considerations, Shimomura revealed his stance on ethics of science and technology.

> To destroy what is constructed by science, we need to further depend on science. Harms and negative impacts caused by science should be weathered by promoting science rather than stifling the development of science. Meanwhile, can we attribute those harms and impacts to science? Aren't these due to our reason's failure to fully catch up with the advancement of science? … What we need to do above all things is to understand the nature of science. This is definitely a philosophical matter... Science should be understood as historical and

societal stuff which has been developed through embracing subjective and objective elements. (Shimomura 1975: 257–261)

At the symposium "Overcoming the Modern" held in July 1942, he emphasised this idea defending himself against other participants' naive idealistic or spiritualistic claims about science technology and machine civilisation (Kawakami and Takeuchi 1979: 259–262).

IMAMICHI Tomonobu, on the other hand, explicitly addressed technology. He argued that today's society is penetrated by technology to the extent that we can call our environment a technologically mediated environment, which could be argued to be an elaboration of Watsuji's concept of the environment. However, rather than following Watsuji in boiling down ethics in a technologically mediated environment to traditional ethics, Imamichi claimed that we need a new ethics to cope with the contemporary condition. He constructs a new virtue ethic for the new times.

The necessity of developing appropriate ethics for the current technological society is described in his insightful studies of "eco-ethica" (Imamichi 1990: 141–150). Imamichi stated that the practical syllogism described in Aristotle's *Nicomachean Ethics* must be reconsidered in the modern eco-environment or human habitat, which is composed not only of nature, but of "technological conjunction." The classical form of practical syllogism is as follows.

Major premise: *A* is desirable.
Minor premise: *p*, *q*, *r* and so on realise *A*.
Conclusion: For some reason, I choose *p* as the means to achieve *A*.

Here, the ideal goal is obvious and the minor premise is the horizon of the freedom of choice, the object of which is a means to achieve the goal. This form of practical syllogism remains valid in our individual decisions even today. Due to the rapid progress of technology and the advent of the technological society, however, means are now more important than goals, and the logical structure of the practical syllogism has been reversed.

Major premise: We have means or power *P*.
Minor premise: *P* can realise goals *a*, *b*, *c* and so on.
Conclusion: We choose *a* as the goal of *P* for some reason.

Obvious powerful means exist, including nuclear, electric and electronic technology. Goals attainable using these means are considered analytically, and hence the means control the goals. The means are so powerful that a selection of a particular goal may have considerable influence on society. However, these sorts of means are usually controlled by groups or organisations, not by individuals; the subject in the modern form of practical syllogism described above is "we." This tends to result in a confusion of responsibilities for goal setting. The problem here is the nosism of organisations, not egoism. Further discussion over Imamichi's thought is found in Chaps. 9 and 10.

Broadly defining technology as a way that allows organisms including living organisms to realise their potential and capabilities to attain their purposes in the real world, SAKAMOTO Hyakudai 坂本百大 (1928-) envisages the establishment of a new philosophy of technology which covers industrial technology, science technology and ICT. He criticised the common behaviour of philosophers that they attempt to understand the nature of technology relying upon abstract or idealistic, but intrinsically inane, concepts such as *nous*, reason or Miki's imagination. He insisted that philosophers now need to be free from such bad habit especially when taking the first step in exploring the nature of technology an individual organism possesses at least potentially, or "micro technology", and the exploration has to be undertaken with incorporating findings from biology and physiology.

On the other hand, technology actualised by groups of organisms, or "macro technology," is non-genetic and thus accumulable, and the accumulated technology becomes an integral element of culture. Therefore, Sakamoto maintained, the essential features of such technology should be captured based on a holistic understanding of human agency, and an assessment technology to correctly evaluate macro technology has to be developed. A prerequisite for this development is to set up standards for assessment. Those who engage in the standard establishment would ultimately be required to address philosophical challenges of, for example, setting goals of the human race or defining values common to all humankind, and the goals and values as the basis of the assessment should be reviewed and revised at every opportunity. These activities are at the core of contemporary philosophy of technology.

Meanwhile, Sakamoto presented a viewpoint that ethics and morals are "deeper-lying" social technology which enables social coordination while characterising ICT as "superficial" social technology. He emphasised the importance of the social technological aspect of ethics and morals in the unstable and uncertain world, and proposed a new research methodology called "ethical engineering" which develops a deductive ethical conceptual system fit in with the reality through a trial and error process of explaining ethical phenomena using ethical concepts (Sakamoto 1972).

Needless to say, representative Japanese philosophers' views of ethics and technology are not exhaustively covered here owing to limited space. Yuasa's thought on technique and ethics is not mentioned in this section, because it is discussed at length in Chap. 6. There are many relevant Japanese philosophers whose names and thought are not described here. Simultaneously, ideas revealed by Miki, Tosaka, Nishitani, Shimomura and Sakamoto are not well incorporated in the discussion in the following chapters. These are subjects of future investigation. On the other hand, this volume contains chapters which address ethics and technology based on the thought of MINAKATA Kumagusu (Chap. 5) and TANAKA Shōzō 田中正造 (1841–1903) (Chap. 12). They are usually not recognised as philosophers, but their words and deeds provide profound insights and pregnant suggestions to those who engage in the research subject of Japanese ethics and technology.

5 Introduction to Chapters

This book is composed of two parts. Part I "Technology from Japanese Philosophical and Ethical Viewpoints" comprises five chapters. The uniting characteristics of this part is that the authors focus on one particular Japanese philosopher and then explores how his ethical thoughts can be applied to technology. The philosophers and thinkers who are represented in this part are NISHIDA Kitarō, WATSUJI Tetsurō, MINAKATA Kumagusu, and YUASA Yasuo.

SHIRAI Masato 白井雅人's chapter is about the philosophy of technology of Nishida. Shirai argues that for Nishida, technology is fundamentally human. He explores the creative nature of technology, that technology is always influenced by traditions, history, and environment but also changes these. He explores the crisis of humanism which is caused by a lack of understanding of the interrelation between subjects and *milieu* in the modern era, and how we can overcome it by taking "thou", the excluded Other, seriously.

INUTSUKA Yū 犬塚悠 revisits the work of Watsuji and, using his ideas, critiques the work of Peter-Paul VERBEEK, a contemporary scholar within the ethics of technology. Verbeek is known to situate morality not only within human decision-making but also in artefacts in themselves. Still, Inutsuka senses an underlying predilection with human decision-making, and the fact that Verbeek suggests Constructive Ethical Technology Assessment, the interaction between a variety of stakeholders in a design process, this human element is confirmed. Inutsuka is inspired by Watsuji to sketch an understanding of good and evil based not on autonomous individuals, but on the concepts of truth and trust within human relationships.

Thomas Taro LENNERFORS also explores the thought of Watsuji. He discusses how Watsuji was interested in technology in his writings, exploring everything from cars to the postal system, claiming that all technologies manifest betweenness. Lennerfors develops Watsujian ethics of technology which is normatively based on the extent to which technologies promote betweenness. He then analyses a contemporary technological system – containerization – which he interprets both materially, as containerized global trade, and subjectively, as the containerization of subjectivities.

SATO Maki 佐藤麻貴 explores the view on science, technology, ethics, and religion by Kumagusu, who was a naturalist, biologist and folklorist. He developed what was later called the Minakata mandala to go beyond the hegemonic scientific worldview – centered around causality, a linear concept of time, and a strong interest in materiality. In contrast, Kumagusu proposes a concept of co-origination, a networked concept of time, and a tripartite division of the world into the material, the subjective, and the event. The event is the unknowable knowledge beneath scientific knowledge. This philosophical view is applied to understand the ethics of implementing large scale technology. Given Kumagusu's view, can we trust scientific-technological solutions to the environmental problems in the world? Can we trust science to capture what is really happening? Most of the consequences of such large scale effort belong to the realm of the unknown, and Kumagusu's ethics implies the necessity to be sensitive to it.

SATO Koetsu 佐藤貢悦, who learnt directly from Yuasa, explains Yuasa's view on ethics and technology. Combining a theoretical and biographical understanding of Yuasa, Koetsu guides us to Yuasa's thought and how it relates to his life and career. As is argued, Yuasa had an intricate relationship to Western philosophy, but aimed to transcend the dominance that *theoria* had over *praxis* in Western thought, which also implied a low status for technicians and technology. Rather Yuasa, inspired not only by Watsuji but also by Carl JUNG, proposed a view where *theoria* and *praxis* are intermingled and where technique plays a central role for ethical development.

Six chapters form part II: "Japanese Ethics in a Technological World". In this part, the point of departure is rather a technological field or issue that is explored by means of the thinking of Japanese philosophers or based in Japanese ethical practice. The technological fields described range from mining to various perspectives on ICT all the way to robots.

MURATA Kiyoshi 村田潔 explains the often amoral stance towards technology and technological development that was described in the beginning of this introduction, particularly in the field of ICT. He argues that this view of technology as purely instrumental and devoid of ethical meanings is due to a withering away of traditional ethical values. This hinders efficient and proactive measures to combat unethical or even criminal online practices. To provide a renewed base for Japanese ethics for ICT professionals, Murata turns to the traditional vocational ethics that was developed during the Tokugawa Era. It is argued that although the Japanese must be aware of such ethics, they also need to see its limitations in an increasingly globalised society.

NAKADA Makoto 仲田誠 focuses on the issue of robots and particularly the status of robots in Japan. Through a boundary spanning exercise through Japanese culture, history, and philosophy, Nakada explores and discusses various existing forms of argumentation about the ethical status of robots. How do robots differ from other artefacts? Because they seem alive, because we like to assign names to them, because they could be anthropomorphic? Nakada argues that a roboethics must be based on betweenness and reciprocity, rather than a more "Western" approach which would stress autonomy or responsibility.

Per FORS and Mikael LAAKSOHARJU have read the work of IMAMICHI Tomonobu, an aesthetician and an advocate of "eco-ethica," to cast light upon Sustainable ICT, which could mean either to make ICT more sustainable or to use ICT-solutions to promote a sustainable society. They focus on Imamichi and his concepts the technology mediated environment, the skilled animal, and the reversal of the practical syllogism. Rather than following Imamichi in his critical and sometimes even nostalgic mood, Fors and Laaksoharju explores how the downsides of contemporary society sketched out by Imamichi could be utilized in a constructive way, to promote sustainability. Is the road towards sustainability a matter of intuition, rather than critical, holistic, analytical thinking?

Elin PALM approaches the issue of ICT based migration governance from the perspective of Imamichi's eco-ethica. Drawing on the theories of the technological conjecture, Palm follows Imamichi in his description of a global, technology-mediated, interconnected world. She argues that migration governance produces what Imamichi calls untimely deaths, deaths that could easily have been avoided. She brings forth a timely argument that developed countries are responsible for migration flows in this globally interconnected society and that ICT is complicit in the production of untimely deaths although it could play a more ethically sound role.

Asai Ryoko 浅井亮子 describes the use of social media after the 2011 Tohoku Earthquake, inspired by Japanese philosopher Ichikawa Hiroshi, who sees technology as something which fundamentally alters our experiences and relationships to others. Asai describes how social media contributed to restoring social relations and social capital with a base in traditional Japanese ethical virtues in a situation where other forms of information and communication technologies were lacking. However, Asai also problematises the ethics of using social media, highlighting the mingling of correct information, disinformation and expired information.

Cheung Ching-yuen 張政遠 focuses on ethical practice in Japan, rather than the view of Japanese philosophers. He has a particular interest in interpreting the Fukushima disaster and its aftermath. He interprets Fukushima as being a victim of Japan's modernisation and seeks an ethical role model in Tanaka Shōzō, a social activist and reformer, who developed an ethics of care in response to the technology-intensive Ashio copper mine and its related pollution case. We follow the activism of Tanaka in rallying support for the cause, all the way up to the Emperor. The piece rings of Tanaka's saying: "True civilization neither ruins mountains, pollutes rivers, destroys villages, nor kills people."

References

Abe, Masao. 1990. "Introduction." See Nishida 1990. An Inquiry into the Good, trans. by Masao Abe, and Christopher Ives. New Haven, CT: Yale University Press, vii–xxvi.

Andersson, Dan-Erik and Lennerfors, Thomas Taro (2011). *Etik* [Ethics]. Malmö: Liber.

Buber, Martin. 1996. *I and Thou*. New York: Simon & Schuster.

Calichman, Richard F. 2005. *Contemporary Japanese Thought*. New York: Columbia University Press.

Carter, Robert. 1997. *The Nothingness Beyond God*. New York: Paragon House.

Davis, Bret. 2011. "Nishitani after Nietzsche: From the Death of God to the Great Death of the Will." In *Japanese and Continental Philosophy: Conversations with the Kyoto School*, edited by Bret W. Davis, Brian Schroeder, and Jason Wirth. Bloomington, IN: Indiana University Press, 82-101.

Ellul, Jacques. 1964. *The Technological Society*. New York: Vintage Books, Random House.

Ellul, Jacques. 1983. "The Search for Ethics in a Technicist Society (trans by Dominique Gillot and Carl Mitcham from "Recherche pour une Ethique dans une société technicienne," *Morale et Enseignement*, 7–20)." *Research in Philosophy and Technology* 9: 23–36 (1989).

Feenberg, Andrew. 2010. *Between Reason and Experience: Essays in Technology and Modernity*. Cambridge, MA: MIT Press.

Fukuzawa, Yukichi. 2012. *An Encouragement of Learning*, trans. by David A. Dilworth. Tokyo: Keio University Press.

Heidegger, Martin. 1977. *The Question Concerning Technology and Other Essays*. New York: Harper & Row.

Ihde, Don. 1979. Technics and Praxis: A Philosophy of Technology. Dordrecht: D Reidel Publishing Company.

Imamichi, Tomonobu 今道友信, 1990. *Eco-Ethica* エコエティカ. Tokyo 東京: Kōdansha 講談社.

Joerges, Bernward. 1999. "Do Politics Have Artefact?" Social Studies of Science 29.3: 411-431.

Katsuya, Akitomi. 2014. "On the Possibility of Discussing Technology from the Standpoint of Nishitani Keiji's Religious Philosophy." *Journal of Japanese Philosophy* 2: 57-73.
Kawakami, Tetsutarō 河上徹太郎, and TAKEUCHI Yoshimi 竹内好, 1979. *Overcoming the Modernity* 近代の超克. Tokyo 東京: Fuzambō 冨山房.
Kimoto, Takeshi. 2009. "Tosaka Jun and the Question of Technology." In *Whither Japanese Philosophy? Reflections through Other Eyes*, edited by Takahiro Nakajima. UTCP Booklet 11, Tokyo: University of Tokyo Center for Philosophy, 121–140.
LaFleur, William. 1985. "Editor's Introduction." In *Zen and Western Thought*, by Masao Abe. Hawaii: University of Hawaii Press, xi–xix.
Lam, Wing-keung. 2011. "The Making of "Japanese Philosophy": Nishi Amane, Nakae Chōmin and Nishida Kitarō." In *Whither Japanese Philosophy? III Reflections through Other Eyes* (UTCP Booklet 19), edited by Takahiro Nakajima. Tokyo: UCTP, 69–80.
Latour, Bruno. 1992. "Where Are the Missing Masses? The Sociology of a Few Mundane Artifacts." In *Shaping Technology/Building Society: Studies in Sociotechnical Change*, edited by Wiebe E. Bijker, and John Law. Cambridge, MA: MIT Press, 225–258.
Murata, Junichi. 2003. "Creativity of Technology: An Origin of Modernity?" In *Modernity and Technology*, edited by Thomas J. Misa, Philip Brey, and Andrew Feenberg. Cambridge, MA: MIT Press, 227–253.
Murata, Junichi. 2008. "Creativity of the Historical World – Nishida and the Philosophy of Technology." *Philosophy and the Hisotry of Science* 哲学・科学史論叢 10: 1–12.
Najita, Tetsuo. 1989. "On Culture and Technology in Postmodern Japan." In *Postmodernism and Japan*, Durham, edited by Masao Miyoshi, and Harry D. Harootunian. Durham, NC: Duke University Press, 3-20.
Nishida, Kitarō 西田幾多郎, 1988. *The Complete Works of Nishida Kitarō* 西田幾多郎全集, 19 vols. Tokyo 東京: Iwanami Shoten 岩波書店.
Nishitani, Keiji. 1982. *Religion and Nothingness*. Berkeley, CA: University of California Press.
Nishitani, Keiji 西谷啓治, 1986–1995. *The Collected Writings of Nishitani Keiji* 西谷啓治著作集, 26 vols. Tokyo 東京: Sōbunsha 創文社.
Nye, David. 2006. *Technology Matters – Questions to Live with*. Cambridge, MS: MIT Press.
Sakamoto, Hyakudai 坂本百大, 1972. "A Course towards a New Philosophy of Technology 新しい技術哲学への道." *Philosophy of Science* 科学哲学 5: 143–165.
Satō, Masahide 佐藤正英, 2012. *The History of Ethical Thought in Japan* 日本倫理思想史. Tokyo 東京: University of Tokyo Press 東京大学出版会.
Satō, Masahide 佐藤正英, 2014. *What Is Japanese Thought: Ethics of Existence* 日本の思想とは何か 現存の倫理学. Tokyo 東京: Chikumashobō 筑摩書房.
Shimizu, Masayuki 清水正之, 2014. *The Complete History of Japanese Thought* 日本思想全史. Tokyo 東京: Chikumashobō 筑摩書房.
Shimomura, Toratarō 下村寅太郎, 1975. *The Philosophy of Science History* 科学史の哲学. Tokyo 東京: Hyoronsha 評論社.
Stenson, S. H. 1989. "Beyond Science and Technology to Absolute Emptiness." In *The Religious Philosophy of Nishitani Keiji: Encounter with Emptiness*, edited by Taitetsu Unno. Nanzan Institute for Religion and Culture: Asian Humanities Press, 114–142.
van de Poel, Ibo, and Royakkers, Lambèr. 2011. *Ethics, Technology, and Engineering: An Introduction*. Chichester: Wiley-Blackwell.
Verbeek, Peter-Paul. 2016. "Toward a Theory of Technological Mediation: A Program for Postphenomenological Research." In *Technoscience and Postphenomenology: The Manhattan Papers*, edited by Jan Kyrre Berg O. Friis, and Robert P. Crease. London: Lexington Books, 189–204.
Virilio, P. 2007. *The Original Accident*, Cambridge: Polity Press.
Watsuji, Tetsurō 和辻哲郎, 2007. *Ethics as the Study on Human Beings* 人間の学としての倫理学. Tokyo 東京: Iwanami Shoten 岩波書店.
Watsuji, Tetsurō 和辻哲郎, 2011. *The History of Japanese Ethical Thought*, vol. 1 日本倫理思想史(一), Tokyo 東京: Iwanami Shoten 岩波書店.

Watsuji, Tetsurō和辻哲郎, 2011b. *The History of Japanese Ethical Thought*, vol. 2 日本倫理思想史(二), Tokyo 東京: Iwanami Shoten 岩波書店.
Watsuji, Tetsurō和辻哲郎, 2011c. *The History of Japanese Ethical Thought*, vol. 3 日本倫理思想史(三), Tokyo 東京: Iwanami Shoten 岩波書店.
Watsuji, Tetsurō和辻哲郎, 2012. *The History of Japanese Ethical Thought*, vol. 4 日本倫理思想史(四), Tokyo 東京: Iwanami Shoten 岩波書店.
Winner, Langdon. 1980. "Do Artifacts Have Politics?" *Daedalus* 109.1, Modern Technology: Problem or Opportunity? (Winter, 1980): 121–136.
Wong, Pak-hang. 2013. "Confucian Social Media: An Oxymoron?" *Dao: A Journal of Comparative Philosophy* 12.3: 283-296.

Part I
Technology from Japanese Philosophical and Ethical Viewpoints

Chapter 2
Technology, Artificiality, and Human Beings in the Later Nishida Philosophy

Shirai Masato

1 Introduction

The philosophy of technology was not a mainstream discipline for a long time (Murata 2014: 57). Traditionally, philosophy dealt with ideas such as being, knowledge, truth, the good, and beauty. In contrast, technique and technology were considered a subdivision of natural philosophy or the philosophy of science. However, in the late twentieth century, the question of technology became a critical issue. Whereas living standards have improved with the development of modern technology, such advances have also brought catastrophe. Nuclear power plant accidents have caused great damage and proliferating weaponry has brought mass destruction. In addition, the development of industrialization has caused extensive environmental destruction. In this reality, some philosophers have questioned the essence of technology. Consider, for example, José Ortega y Gasset, Hannah Arendt, Karl Jaspers, and Martin Heidegger. Heidegger in particular has greatly influenced this field, searching for the "essence of technology" and pointing out the problems of modernization and negative aspects of technology. Thus, the problem of technology has become an important issue in the field of philosophy.

Nishida Kitarō 西田幾多郎 (1870–1945) is considered one of Japan's greatest academic philosophers (Heisig et al. 2011: 646). He developed a philosophy of

---

The original version of this chapter was revised: MS Mincho font was updated throughout the book for Japanese and Chinese characters. The correction to this book is available at https://doi.org/10.1007/978-3-319-59027-1_13

Shirai Masato (✉)
Inoue Enryo Research Center, Toyo University, Bunkyo, Tokyo, Japan
e-mail: masshira@drive.ocn.ne.jp

Thomas Taro Lennerfors, Murata Kiyoshi (eds.), *Tetsugaku Companion to Japanese Ethics and Technology*, Tetsugaku Companions to Japanese Philosophy 1,
https://doi.org/10.1007/978-3-319-59027-1_2

technology in his later works,[1] which places technique at the heart of his philosophical system. Nishida characterized humans as creative beings and emphasized the importance of technique for its ability to facilitate creation. He did not address the negative aspects of technology in detail, as did Heidegger, however, his ideas about technology are very suggestive. In this chapter, I will elucidate Nishida's philosophy of technology and argue for its applicability to the modern world.

According to Nishida, technique can exist in harmony with humans and nature. He qualifies this, writing, "However, that man steps over the self itself in the direction of willfulness is the degeneration of man, that reason steps over the self itself in the direction of abstraction is to lose the objectivity of reason (that technology steps over technology in the direction of technology is artificiality)" (9: 58).[2] Nishida thought that subjectivism and anthropocentrism cause artificiality in technology and destroy the harmony between humans and nature. To remedy this, objectivity and humanism in the true sense are required; and here Nishida stressed the importance of considering the root of humanity.[3]

Proceeding from this idea, I will elucidate Nishida's philosophy of technology. In Sect. 2, I outline Nishida's thinking regarding the creation and technique of human beings. Nishida believed that the difference between humans and animals lies in the ability to create a thing or not. In order to create, one must comprehend objective conditions and suitable materials. Therefore, technique is required, and technique connects our activities with nature.

Section 3 discusses the problems of human reason and technology that cause the alienation of human beings from nature. Because reason can observe reality, it pro-

---

[1] On Nishida's philosophy of technology, KOSAKA Kunitsugu 小坂国継 writes: "His theory of technology lacks consideration for historical situations of technology, and he hardly considered the problems of technology which are presented by Spengler, Ortega and Jaspers. Moreover, he hardly considered the technology of politics which unifies all technology" (Kosaka 1995: 175). On the other hand, ŌHASHI Ryosuke 大橋良介 valued the contemporary relevance of Nishida's theory of technology, comparing him with Heidegger, though dismissing issues of nihilism (Ōhashi 1995). MURATA Jun'ichi 村田純一 also pointed to Nishida's philosophy of technology as having contemporary significance in its emphasis on the creativity of technology (Murata 2014). In this chapter, I link Nishida's consideration of history with the problems of technology.

[2] Here and below, quotations from Nishida's works are based on the following edition: SHIMOMURA Toratarō et al., eds. (1965–66), *Complete Works of Nishida Kitarō*, Tokyo: Iwanami Shoten (in Japanese). The location of a quotation is indicated by the volume number followed by the page number. I referenced the following translations, but I have properly corrected them. Translations: Haver, William. (2012); Krummel, W. M. John and Nagatomo, Shigenori. (2012).

[3] Before his later works, Nishida did not argue that the problem of technology was an important issue. He had discussed the problem of the relationship between subject and object as "intellectual intuition." In his first book *An Inquiry into the Good,* he writes: "the spirit which pursues an ideal requires infinite unity; and this unity is given as intellectual intuition" (1: 42). Furthermore, he explains that this "intellectual intuition" is "artistic intuition" (1: 40). Before his later works, intellectual intuition was the ultimate goal, and it was achieved through artistic intuition. In his later works, the problem of the relationship between subject and object was argued as "action-oriented intuition." Action-oriented intuition is not artistic intuition, but connects history and technology. He thought intellectual intuition as artistic intuition was based on action-oriented intuition as historical and technological intuition (7:335). This is the reason why this chapter focuses on Nishida's later works.

duces a subjective position that is removed from reality. Similarly, technology can change objective conditions, and therefore it has a tendency to take a subjective position. As a result, reason and technology lose their connection to objective nature. This lost connection causes a crisis in which humanity appears to be lost.

In Sect. 4, I present the possibility of recovering humanity by discussing the problem of "I and Thou." In order to recover objectivity, one must obey the objective "ought (*Sollen*)." This ought is connected to *agapē*, forcing the question of what is required to love "thou." This "thou (*nanji* 汝)" is not a homologous "you (*kimi* 君)." We encounter "thou" only by hearing the voices of those who are oppressed by our homogeneous society. In this way, we can negate in our society what oppresses "thee" and remake it anew. By encountering others in the true sense, we can recover objectivity in the true sense and thus humanity.

2 Creativity and Technique

Nishida characterized humans by distinguishing them from animals. He wrote:

> While human beings have been said to be ζῷον πολιτικόν [social animals] or ζῷον λόγον ἔχον [rational animals] or as sensible or rational, we are rather—as [Benjamin] Franklin stated—tool-making animals. But tools are not made for the sake of themselves. Making tools must instead entail making things. Things (*mono* 物) refer to objectives (*kyakkanteki* 客観的)[4] as well as universal things, things to which we cannot do anything, and things that change for themselves. Animals make things as well, and beavers are said to be ingenious architects. And certain animals are even thought to make tools. But it is inconceivable that an animal looks at a thing as a thing, [for] animals do not possess a world of objects (*taishōteki* 対象的). (8: 276–277)

As this quotation makes clear, the important points distinguishing humans and animals are humans' creative ability in making tools and observing things as objects apart from the self. Animals also see things, but they only look at them as subjects for appetite, sexual desire, and threats to their lives. Their action "probably does not extend beyond its bodily movement" (8: 277). Nishida said, "Even if we say that animals possess tools to a certain degree, we cannot say that they truly possess tools. Tools must be substitutable" (8: 277). For example, apes use a branch for reaching distant objects, but they use a branch only as an extension of their hands. This means that animals cannot improve tools and produce them in large quantities. On the other hand, humans make tools on a large scale and replace them when they are damaged. Humans see a thing as a thing and make tools that are broadly shared.

[4] In later Nishida Philosophy, the word "objective" in a broad sense means something that places restrictions on our actions. For example, when we see a stone on the road, we avoid it by walking around it. This stone is the objective thing that affects our action. As physical laws affect our action, we call them objective conditions and as the "ought" has a claim to our defense, it is objective ethical law. In a narrow sense, "objective" means "unchangeable." We cannot change physical laws; they have objectivity in a narrow sense. In the same manner, the "ought" has objectivity in a narrow sense, because we should not change it selfishly. On other hand, we can change a stone into sand by using a stonebreaker machine. The stone, therefore, does not have objectivity in a narrow sense.

As this shows, seeing a thing and making a thing are the most important elements of human action. Thinking they cannot be separated, Nishida constructed one concept named "action-oriented intuition" (*kōi-teki chokkan* 行為的直観), which is worth exploring. Action-oriented intuition is defined as "acting is seeing and seeing is acting." This can be considered in terms of our everyday experience.[5] Our actions always become possible in conjunction with the act of seeing (intuition). Nishida explained this with the example of a carpenter. "Aristotle said,[6] 'The form (εἶδος) is in our mind,' but the form of a house is not only in a carpenter's mind, because the form of a house is not just fantasy. It must have objective possibility. It can be constructed with objective conditions and technology (*gijutsu* 技術)" (7: 338). To build a house, the carpenter first needs to see the site where the house will be built. The plans for the house are drawn up with due attention paid to objective conditions such as sunlight, drainage, topography, and climate. Building materials suited to these objective conditions are then selected and the house is built. It is through seeing that the act of building a house becomes possible. At the same time, this "seeing" is impossible without the act of building the house. It was precisely because we began building a house that we saw the objective conditions of the site and saw suitable building materials. If one were a geologist, one would see the site differently from a carpenter: one would see strata and geological features, not sunlight or drainage. We see things in different ways depending on what we make. In this sense, "seeing" became possible in the act of making. Furthermore, through the act of building a house we came to see the built house. "Acting is seeing and seeing is acting" thus means it becomes possible to make something through seeing, and that thing is seen through the act of making it.[7]

As mentioned above, Nishida used the term "technology" in relation to action-oriented intuition. According to him, technique is required when we make a thing through seeing.[8] He wrote, "Technique is not something that merely appertains to the [epistemological] subject. It means that our self enters into the thing, and that activity of the thing becomes in harmony with our activity" (8: 297). Naturally, the phrase "enters into the thing" is not meant in the literal sense; instead, it refers to activities of ourselves, and the things do not hinder each other. For example, when

[5] SUZUKI Daisetz 鈴木大拙 argues that action-oriented intuition is undifferentiated cognition (*mufunbetsuchi* 無分別智) (Suzuki 1989: 167). It is his characteristic interpretation, but perhaps it leans toward Buddhism.

[6] Although Nishida referred to Aristotle, according to a footnote to the *New Series of Complete Works of Nishida Kitarō* by KOSAKA Kunitsugu, the same words are not found in Aristotle's works. Related descriptions are found in 1032b1 and 1070a8–16.

[7] I have discussed the subject in this paragraph elsewhere, see Shirai (2013: 106–7). For more details on action-oriented intuition, see also Kosaka (1995: 69–116) and Itabashi (2008: 15–38). In English, Robert E. Carter discusses "action-oriented intuition" (Carter 1997: 105–110); however, he stressed the religious aspect of action-oriented intuition and does not take up the question of making a thing.

[8] Nishida uses the term "*gijutsu*" without distinction between technology and technique because the Japanese word "*gijutsu*" means both technology and technique. In this chapter, I use the term "technology" for mass-scale and high-level techniques.

we try to make a wooden statue, we should do so in accordance with the nature of the wood. If we violate it, the wood will break and we cannot complete the statue. Thus, Nishida said, "We cannot accomplish anything unless it is thoroughly through nature's technique" (8: 298). We cannot make anything if we go against nature. He also wrote, "The distinction between an animal species and a human society lies in which our self transcends our self and comes to be in harmony with expressive nature" (8: 294). The term "expressive (hyōgen-teki 表現的)" means that nature appears in different ways based on what man wants to do. For example, a tomato appears as an ingredient for a cook, but as a motif for a painter. Being expressive takes on various appearances.

It might be argued, however, that building a house is a special experience and not an everyday experience. Although it is true that Nishida does not specifically say so, "walking" is probably also characterized by action-oriented intuition. For example, when walking, we look at the state of the road's surface, the state of traffic congestion, the state of the weather, and so on. Moreover, we can never walk when we try to walk against our physical conditions or physical laws. We need a technique for walking that is in harmony with nature. What is the relationship between "walking" and "making something"? "Walking" has an aim; through walking, man arrives at his destination. Even when the aim of walking is walking itself, the healthy self would be made. Based on the above arguments, it is clear that most daily experiences such as "walking" are also actions such as "action-oriented intuition."

"Action-oriented intuition" has one more feature, which Nishida considered a movement "from created to creating (*tsukurareta mono kara tsukuru mono he* 作られたものから作るものへ)." "Creating" something starts from something that has been "created." When we began building a house, the idea of this house did not come from a vacuum. People traditionally build houses when they settle in a certain locality, and it was in the context of this tradition that we thought of building a house. For instance, nomads move together with their animals and erect tents instead of building houses. Our act of "making" (i.e., building a house) has been defined by society and history. Consequently, when making something, we start out from a socially and historically "created" state. At the same time, we build a new house that has not previously existed. Referring to different kinds of houses built in the past, we build a new one incorporating our own new ideas and new technologies. In this sense, we shift from "that which has been created" to "that which creates" and add something new to society and history.[9]

Let us summarize the details of Nishida's "action-oriented intuition." A human being is characterized as a creative being who makes a tool and makes a thing using tools. When one makes a thing, one sees a thing objectively. Seeing a thing objectively means one sees objective conditions and suitable materials. Here, technique is required, and technique connects our activity with nature. In the next section, I will discuss the problems concerning modern technology.

[9] I have also discussed the subject in this paragraph elsewhere, see Shirai (2013: 107).

3 Human Reason and Technology

Nishida argued that human reason was connected with the "historical world." When we create something, we create a new historical world at the same time. Let us examine this historical world with reference to Nishida's own words.[10]

Nishida wrote that "In the historical world the subject (*shutai* 主体) determines the *milieu* (*kankyō* 環境) and the *milieu* determines the subject" (9: 50–51). Here, "*milieu*" does not mean only the natural environment, but signifies the historical context as a whole, including the natural world to which human beings belong. "Subject" refers to human beings. The statement "the *milieu* determines the subject" means that human beings are defined by the historical conditions in which they find themselves, while "the subject determines the *milieu*" means that at the same time, human beings create new conditions. Nishida continues: "While subject and *milieu* are in utter mutual conflict, the subject determines the *milieu* by individualistically negating its self, and the *milieu* determines the subject by individualistically negating its self" (9: 51). While being defined by historical conditions, human beings, by exercising their own individuality, negate the self that is merely defined by historical conditions and create a new thing and new self. This creation of a new self also leads to the changing of the *milieu*. At the same time, historical conditions, while having been created by human endeavors, go on to define human beings living in that period because the created conditions turn into the individuality of that historical period.

For example, when a carpenter plans to build a house in Japan, they would build a wooden house that adheres to Japanese traditions and differs from a tradition of building brick houses. For solving the problem of humidity peculiar to Japan, a well-ventilated house is built. The carpenter's idea of a house is determined by the *milieu* that includes historical and spatial conditions. This means "the *milieu* determines the subject." However, the new house reflects the carpenter's individuality; they add a new idea to the work and build a new house that never existed before. They never repeat old ways and they create new styles. In this fashion, carpenters negate the old forms of 'house' and their past selves. The tradition is renewed by building a new house and many other new attempts. This renewed tradition is a new *milieu* which is determined by the subject. At this time, the subject negates the old *milieu* and one's past self by exercising one's individuality. This means "the subject determines the *milieu* by individualistically negating its self." We negate old tradition and our past selves by exercising our individuality. Furthermore, we build another house based on the renewed tradition. This renewed tradition has the individuality of a new epoch, while it also belongs to one traditional line. Although we belong to one traditional line, we build different houses in each epoch. Each epoch has its own characteristics and we create different things according to the characteristics of each epoch. New epochs negate old epochs and each has individuality. This

[10] I have also discussed Nishida's understanding of human reason briefly elsewhere see Shirai (2014: 136–8). In this section, I have developed the discussion in more detail.

means "the *milieu* determines the subject by individualistically negating itself." In this fashion, by exercising their respective individualities, both human beings and their *milieus* come to define each other, while negating the given mode of being of their respective selves. Nishida therefore concludes that "Subject and *milieu*, through their individuality, determine each other through mutual negation; moving from the created to the creating, the world continues to determine itself individualistically" (9: 51). By mutually exercising their individuality, human beings as subjects and the *milieu* as historical conditions determine each other. In this way, the world moves from created historical conditions to one that creates new historical conditions. Furthermore, as something possessing an individuality that cannot be reduced to the past or future, these historical conditions become a locus for new creation.

From the above, it is clear that individuality is closely connected with freedom and creativity. We create new things with our own new ideas free from *milieu* and surpassing our old ideas. However, it does not mean being isolated. While we create a new thing with the help of *milieu* and tradition, we add new ideas to it. Exercising individuality means freedom from the constraints of the past and creating something new. In the same manner, negation is closely connected with freedom and creativity. Usually, negation has a negative meaning in Japanese as well as in English. However, Nishida uses this term in a positive way. Through negating *milieu* and tradition, we become free from them and create a new thing. The negation is not a mere denial. We negate the past and create a new history.

Nishida characterizes this historical movement as something "mediated by absolute negation (*zettai hitei* 絶対否定), mediated by the absolutely transcendental (*zettai-ni chōetsu-teki narumono* 絶対に超越的なるもの), by absolute nothingness (*zettai mu* 絶対無)" (9: 51). Why does he use the expression "mediated by absolute negation"? Perhaps it is because he thinks of the historical movement as neither something causal that results from past causes, nor something teleological that is defined by a future goal. Not being defined by causes or goals, the given reality is negated and a new reality is created. This world is not something that is completely defined by causes and goals. Rather, it is a locus for unrestricted, free creation in which the given reality of the past is negated and a new reality is created. Because our individuality is only one, by exercising our individuality we create a new thing that has never before existed. It was this aspect of the negation of reality that Nishida expressed as being "mediated by absolute negation."[11] If this were not the case, our actions would be defined by the past or the future, and it would be impossible for us to become free and responsible subjects. Because we pass the responsibility of our act to past causes or future goals, we say "It is not my will, it

[11] On absolute negation, Krummel writes, "Absolute negation is the movement of the absolute nothing. Delimited by nothing, the universal must ultimately be nothing, no determinate universal" (Krummel 2015: 101). Although he does not connect absolute negation with historical movement, his assertion can be applied to historical movement as follows: history is delimited by nothing and it is not delimited by the causal past and future goals.

is determined by the past or the future." Insofar as we are free and responsible subjects, this world must be mediated by absolute negation.[12]

Why is the character of this negation described as "absolutely transcendental"? This is because the power to negate this reality does not lie in the given reality as a past cause or future goal. If the power to negate reality lay within the given reality, be it a past cause or a future goal, one would end up positing some form of continuous current of history rather than a new creation negating reality. Consequently, the world would become subject to either the past or the future and could not become a locus for new creation. It is precisely because the mediating agency of negation cannot be discovered within the given reality that it turns into the power to renew the reality. We create the new thing beyond the given reality and renew the world by negating past causes and future goals. Therefore, it is called "absolutely transcendental". And because it never manifests itself in the given reality, it is also called "absolute nothingness." Like the word "negation," "nothingness" has a negative meaning in Japanese as well as in English. However, Nishida also used the term "nothingness" with a positive meaning. Absolute nothingness means that there is nothing to disturb our action. "Mediated by absolute nothingness" is the condition where we are mediated neither by the past nor by the future. Through the negation anything that restrains us comes to nothing, we create new things freely with no barriers. This does not mean that nothingness exists, but that we come to be free and creative through nothingness.

It is in this mode of being of the world that Nishida situates reason. He writes: "Because it is mediated by absolute negation, the world moves with self-awareness (*jikaku* 自覚). This is to say that it is characterized by reason" (9: 51). He also refers to "the center of the individual structure whereby the world determines itself, that is, reason" (9: 51). Being mediated by absolute negation, the present, defined by neither past nor future, is created as a unique present possessing its own individuality. It is in this creation of something possessing its own individuality that Nishida sees the workings of reason. For Nishida, reason signifies the power that through the mediation of absolute negation creates history. Of course, reason was also associated with universal concepts and laws. Being an individual structure is not contradictory to universality. Because this individuality is gained by absolute negation, it includes universality as the negation of individuality. Specifically, to create something we must look at objective conditions containing physical laws, among other things. Thus, it can be said that we refer to a universal concept for creating something new. Nishida also says of reason that it "moves with self-awareness." It is here said to act with self-awareness because by individualistically creating the world, one becomes cognizant of one's own individuality and objective world.[13] By being mediated by absolute negation, one discovers one's own individual possibilities instead of being defined by other possibilities of the past or future. Therefore, reason is the act of being self-aware while creating history. To put it briefly, reason is

[12] On free will in Nishida's philosophy, see Taguchi (2006).

[13] For a detailed discussion of the scientific recognition, see Shirai (2006: 108–9). On general research about epistemological problems in later Nishida philosophy, see Itabashi (2008).

the action through which one creates an individual world through recognition of a universal concept.

Due to the nature of reason as characterized above, Nishida's diagnosis of the period in which he lived becomes possible. I wish to examine in more detail the relationship between reason and a particular period in history.

As discussed above, the historical world, with absolute negation as a mediating agency, constitutes a movement that negates itself to create a new reality. This world comes to possess the characteristic of "always transcending the self itself; the historical present is always vacillating" (9: 57). In other words, because it possesses the power to negate reality, it transcends the given reality and not having any fixed orientation, it vacillates between various possibilities. Consequently,

> [T]herein is the subjectivity of the world. Therefore, as the apex of that which is created and creates, man is always willful. Where reason steps over reason in the direction of reason is where the world of abstract logic is established. To see the world from such a position is subjectivism. (9: 57)

To transcend the given reality means to produce a subjectivistic position viewing reality from a position that is removed from reality. Moreover, not being constrained by reality makes it possible for man to behave in a willful or arbitrary manner. Furthermore, by creating a new reality while negating reality, reason goes on to create the world of abstract logic through being completely removed from reality. In other words, the historical world is intrinsically endowed with the possibility of removing itself from reality and moving to a willful and abstract position.[14] Nishida continues:

> However, that man steps over the self itself in the direction of willfulness is the degeneration of man; that reason steps over the self itself in the direction of abstraction is to lose the objectivity of reason (that technology steps over technology in the direction of technology is artificiality)… The world of the historical present that is always transcending the self itself persistently proceeds in the direction of transcending the self itself. It is the direction of individualization and progress, and it is also the direction of degeneration, the furthest point of which is decadence. A single historical tendency cannot go beyond this. (9: 58)

To be willful is the essence of a human being, but to merely behave willfully as one pleases means that one's humanity is lost and one becomes degenerate. Again, to take abstract reason to extreme lengths results in thoroughgoing subjectivism removed from reality and in the loss of objectivity. The transcending of reality by reality means progress, but it also means the loss of humanity and objectivity through the aggrandizement of subjectivity, namely, that man intrinsically contains within himself a disposition towards decadence. As a result, he reaches an impasse beyond which he cannot go. Moreover, this is a possibility with which reason and man are intrinsically endowed.

---

[14] For reasons of space, I cannot cover the questions of willfulness, freedom, and inevitability in Nishida's philosophy. For a discussion of these questions, see Itabashi (2008: 226–257). For Nishida, freedom and inevitability are inseparable, and inevitability and freedom are established in the acceptance of historical inevitability within freedom. In this sense, it is said that "freedom is no different from inevitability, and inevitability is no different from freedom."

In the progress of subjectivism and loss of humanity, technology falls into artificiality. As mentioned above, technique essentially is based on looking at things and brings harmony between humans and nature. Technology as mass-scale and high level technique also consisted of harmony between human and nature. However, in thoroughgoing subjectivism, subject separates from objective things and nature. Thus, technology lost its purpose as harmony between them and became an instrument for the expansion of subject. As a result, technology lost its objectivity and degenerated into technology for its own sake. Furthermore, technology lost humanity and oppressed humans because subjectivism overlooked the human objective condition and put it at risk.

These possibilities have become a reality in the impasse reached by man in the modern age. Let us next consider Nishida's observations on how this impasse of the modern age arose.

After having briefly touched on Greek culture, Nishida writes of European culture in the Middle Ages:

> God and man were opposed; the immanent and the transcendental were in utter opposition. So it sought man's source in the transcendental; that is to say, it was a religious culture… However, one can say that in one aspect God was made human. And that is the secularization of the Church… On the other hand, anything truly religious was lost. (9: 60)

Nishida did not consider the Middle Ages as a truly religious culture. In that time, after which people regarded God as too transcendental, the relationship between God and humans was completely divided. The Church was also disconnected from God and came to merely represent a human society. Despite its role of mediating between God and humans, the Church found itself in opposition to God and became secularized. It was a society in which the Church had become a center of power and had been secularized, representing a world ruled by a God made human. As for the Renaissance that arose after the Middle Ages,

> [It] was not merely the restoration of ancient culture but the fact, it is said, that man discovered man. Man returned to the creative self and wrested sovereignty from God. This was humanism. And one can probably say that it was there that the modern culture of anthropocentrism began. (9: 60–61)

As a movement that restored sovereignty from God to man, the Renaissance set the direction of anthropocentrism of the modern age. In this fashion, the modern age shaped a magnificent culture, "but the development of anthropocentrism had of its own accord to progress in the direction of subjectivism and individualism. Reason stepped over reason in the direction of reason. There man conversely lost man himself" (9: 61). The anthropocentrism of the modern age ended up progressing in the direction of subjectivism and individualism, resulting in reason becoming abstract and losing its objectivity, with humanity too being lost.

Nishida described this process of the loss of humanity in terms of a conflict between nature and man. During the Middle Ages, God and man had been in conflict, but the modern age became a period of conflict between man and nature. According to Nishida, "Nature, being *milieu*, was even something to be used, but nature-as-object must essentially be something that negates man… Man, at his beginning and at his end, has no choice but to be negated by nature" (9: 61). For subjectivized human beings, nature has become an object opposite from subject. The more expansive the subject was, the more serious was the conflict between subject and object. For human beings, nature started to be observed and utilized as a resource. In a conflict between nature and man, technology loses the harmony between them. Through technology, subjectivized man uses nature for himself and tries to change nature as he wishes. However, at the same time, nature repudiates man. As indicated by the statement that "Man, at his beginning and at his end, has no choice but to be negated by nature," human beings can control neither their birth nor their death. Birth and death, as examples of the providence of nature, repudiate man's seeking to control everything. Furthermore, "Even though we speak of conquering nature, it is only because we conform to nature that we conquer nature. Our hands and feet alike are things. Even that which may be considered to be inner desire negates the ego" (9: 61–62). In other words, although we may say that man conquers nature, this is possible only by obeying the laws of nature, and even our hands and feet are no more than "things" obeying the laws of nature. For subjective[15] human beings, even their hands and feet have elements rejecting human control. Again, even the inner desires of human beings arise as natural instincts and reject control by subjective human beings. It is thus concluded that "There is no way of discovering the self in nature-as-object. Therein there is only death. It is for this reason that I say that anthropocentrism conversely leads to the loss of humanity" (9: 62). One can find oneself neither in nature outside oneself, nor in one's body, and not even in one's inner desires. For a subjective ego, birth and death, objective nature, and body and desire cannot be controlled. Everywhere there is only that which rejects the self, and wherever one looks, there is only death of the self.

In this fashion, subjectivism *qua* anthropocentrism reaches a dead end and experiences its own destruction. Technology reached the same fate. Nishida wrote, "that technology steps over technology in the direction of technology is artificiality." Technology loses its objectivity and humanity. Furthermore, technology becomes technology for its own sake and oppresses humans. Let us give a concrete example.

While the technology of touch screens increases, the blind face difficulties. On January 8th, 2009, an electronic version of Reuters distributed an article entitled "Touch-screen gadgets alienate blind" (Carew 2009). It said "With the popularity of touch screens, once simple products such as televisions and stereos have become difficult for blind people to use as they often require navigation of multiple menus that need to be seen to be used effectively." It is impossible for blind people to read information on the flat panel of a touch screen. In the Japanese Diet, the problem of touch screens was also raised as serious issue (House of Representatives 2004: 16–17). According to the proceedings, the introduction of touch-screen ATMs

[15] The word "subjective" in this chapter means something that projects an action and observes an object. Sometimes our hands and feet do not move in accordance with our intentions. We need to observe and understand our body for an action to be intended. Projecting and observing is an important factor for understanding the word "subjective."

(automated teller machines) has created difficulties for many blind people. Although they could use the now-obsolete ATMs, they cannot use the new touch-screen models. New technology alienates blind people from society.

This means that technology has the possibility to oppress humans, especially minority groups. Subjectivism loses objectivity, and at the same time erases those others who have different characteristics. The subject is homogenized for the same reason and the objective disparity of bodies is eliminated. Thus, subjectivized technology imposes a burden on people who belong to minority groups. In the case of touch screens, nobody noticed the problem at the stage of development and of implementation. When old ATMs were replaced by touch-screen ATMs and blind people started to point out the problem, we finally became aware of it. We cannot realize the problem by ourselves. In order to recognize the problem, we must hear the voices of those who are oppressed. In the next section, I will discuss the problem concerning relationships with other people.

4 Humanity and "I and Thou"

As discussed in the previous section, our life has reached a dead end in the modern age. Nishida writes that to break free from this impasse, "We must again return to the bottom of creative nature. From there we come to regain a new creative power, and new life is born in us" (9: 58). Through the creative power lying at the foundations of ourselves, we will break free from our impasse and gain new life. Furthermore, this creative power, with absolute negation as its mediating agency, is something transcendental. Thus, Nishida also writes:

> What is called our true self exists in being productive and creative as a creative element of the creative world (*sōzō-teki sekai no sōzō-teki youso* 創造的世界の創造的要素). Therefore, man exists in the fact that what is truly immanent in the self itself is mediated by the transcendental and what is mediated by the transcendental is truly immanent in the self itself (9: 62).

To gain new life as a creative self, transcendental creative power must simultaneously become power that is immanent in ourselves. In the "contradictory self-identity (*mujun-teki jiko-dōitsu* 矛盾的自己同一)"[16] in which what is transcendental for us is simultaneously immanent in us, it becomes possible for us to live a new life. Thus, Nishida concludes:

> But when I speak of returning to the transcendental, I am not speaking of merely returning to an abstract absolute being that is the negation of man. It is to take up the position of truly individual historical reality; it is to take up the position of historical reason. It is to return to the position of the self-identity of absolute contradictories that establishes a world that, because it is transcendental, is truly immanent. (9: 65–66)

[16] For more details, cf. Kosaka (1995: 182–216) and Itabashi (2008: 185–314).

Nishida says of this contradictory self-identity that, "From the position of the man of religion, this contradictory self-identity will be what is conceived of as God" (9: 66). In other words, from the position of a man of religion, to return to the position of contradictory self-identity means returning to God, but it does not mean returning to an absolute being conceived in abstract terms. Rather, it means to take up the position of individual historical reality. How then does it become possible to take this position? Nishida writes the following in connection with the historical world as a world transcending ethnic society:

> When the global world is formed, it might be thought that the world loses all its various traditions and becomes non-individual, abstract, universal, antireligious, and scientific. This was the direction of modern Europe's progress. As the self-negation *qua* affirmation of absolute being, such a negative aspect must be contained from the outset in the direction of the self-formation of the global world. An aspect of negation of the human is contained in the historical world. (11: 457)

Here too, Nishida states that the inevitable outcome of modern Europe will be rejection of the human. However, this rejection of the human as the future of modern culture cannot be overcome by simply negating the modern age, for as we have already seen, culmination in rejection of the human is an inevitable consequence of history. Therefore, although "People concerned about the future of European culture often advocate a return to the Middle Ages (like Dawson)," Nishida is critical of this viewpoint: "Modern culture has developed through historical inevitability from the culture of the Middle Ages. Not only is it impossible to return to the position of the culture of the Middle Ages, neither is it a means to save modern culture" (11: 460). We can now also understand Nishida's following criticism:

> While I am by and large in agreement with Berdyaev's "*meaning of history,*" his philosophy does not go beyond Böhmian mysticism. The new age must above all be scientific. Tillich's "*kairos and logos*" also have points in common with my epistemology, but his logic is unclear. Today these new trends must be given a thoroughly logical foundation. (11: 463)

Proceeding in a reactionary manner toward mystical, unscientific, and illogical discourse will not save the current age from its crisis. How then can humankind be rescued from this path of rejection of the human? How can we achieve both returning to God and being scientific? It is possible by recovering objectivity because the cause of problems is subjectivism. However, this "objectivity" never means the opposite of subjectivity. As discussed above, in the conflict of subject and object, there is no way to save humanity. Nishida said, "We confront absolutely objective expression, it can be thought that we then confront, as it were, the word of God (*kami no kotoba* 神の言葉)" (9: 21). He also said, "Objective expression is thought as an infinite 'ought (sollen)'" (9: 22).[17] According to Nishida, the "ought" as the word of God is objective in a true sense.[18] Concerning the ought, Nishida wrote, "The ought as the objective expression of the historical world, that itself forms itself

[17] For a detailed discussion of what ought to appear in the historical world, see Shirai (2007). On the relationship between tasks and the absolute, see Sugimoto (2013: 204–5).

[18] For a detailed discussion of what ought to be considered as word of God, see Shirai (2008).

in expressive activity, confronts us from without" (9: 54). This "from without" means it is beyond the realm that we can change. The objectivity of the ought is based on the fact that the subject cannot change it. However, unchangeability alone is not enough. Nishida said, "As acting-self we confront absolutely infinite objective expression (the so-called word of God), we become self-aware[19]; that is, we become personal (*jinkaku-teki* 人格的)" (9: 47), because we become a personal-self through the ought. Even though the ought is objective, the ought is not in conflict with the subject in the sense that we become ourselves by obeying the ought.

What then are the specific contents of the ought? Nishida wrote, "By confronting the utterly transcendent one, we become true personalities. And the fact that the self is the self by confronting the transcendent one means at the same time that I confront my neighbors with *agapē*" (9: 216). Nishida similarly writes: "We are a personalistic self by loving neighbors like ourselves in imitation of God's *agapē*" (6: 424).[20] These words remind one of scenes in which Jesus Christ preached love for our neighbors, especially the following words: "A new commandment I give unto you, That ye love one another; as I have loved you, that ye also love one another" (John 13: 34).

It is noteworthy that when Jesus spoke of love for our neighbors, he was speaking of outsiders such as Samaritans who were not usually considered as neighbors (Luke 10: 25–37), and he meant, "Love your enemies… and pray for them which… persecute you" (Matthew 5: 44). As Katsumi TAKIZAWA pointed out,[21] Nishida clearly distinguished "I and thou (*watashi to nanji* 私と汝)" from "I and you (*boku to kimi* 僕と君)." In Nishida's terminology, the "you" is grasped through its homogeneity with me, and "is merely another I and only just a neighbor (not 'thou' in the true sense)" (8: 69). We are required to love those who, when considered from the position of ourselves, cannot possibly be described as neighbors. To "confront my neighbors with *agapē*" means to "imitate God's *agapē*." As Jesus Christ died on the cross for sinful humans, God's *agapē* was to establish us through his self-negation. Therefore, to confront our neighbors with *agapē* in imitation of God must mean to bring out the best in others through our own self-negation. At this time, we should negate not only our own self but also our historical scheme which determines our self. If not so, we cannot say it is self-negation in the true sense. To encounter "thee" means to realize that the historical schema continuing from the past that defines us is violently pushing "thee" into the position of enemies and outsiders, and to bring out the best in our neighbors with love. It is required that we build relationships with others by becoming aware of and negating the schema of society, which violently assumes others to be outsiders and enemies.

---

[19] For a detailed discussion of self-awareness, see Shirai (2006, 2013). Further, for a discussion of self-awareness with a focus on "reflecting," see Okada (2001).

[20] I have discussed Nishida's understanding of the *agape* of God as works of God briefly, see Shirai (2015: 111–112). In this section, I develop its discussion in more detail.

[21] Cf. Takizawa (1972: 150).

Negating the schema of society in this way includes both negating and transforming a historical society, or a "historical species (*rekishi-teki shu* 歴史的種)"[22] in Nishida's terminology. Nishida wrote: "We are born in a species, but we must be beings forming the species" (8: 450), and the individual "is even considered to rupture the species in some cases" (8: 450). We are born in a historical society as a historical species. At the same time, we go beyond a specific society and should transform it. Nishida wrote:

> Reality is reality as long as it goes beyond itself, and out of itself. Life is a life as long as it is creative, contradictory, has self-identification, and has a chance to break form, even though it has the fixed form of a species. (8: 451)

Of course, negation of a species does not make us depart from the species. It means acting to make a rigid society come alive. The true transformation of a species means that we must always be creative. Nishida wrote:

> Our life as belonging to a species is also the result of an infinite dialectic development. If we act conventionally as a species, it is a mechanization of self and death of the species. We must be creative from moment to moment. (9: 190)

We are compelled to be creative. We are forced to live in reality while facing the negation and forming the reality anew. These tasks demand the transformation of the historical species to which we belong. It only becomes possible for us to encounter "thou" in its truest sense when we break through the homogeneity implied in "I and you." We encounter "thee" not as the homologous other, but as the absolute other by discovering our true task in our society and breaking free from society. Once again, we can only encounter others in the truest sense by negating ourselves and our society and achieving a clear recognition of the ought through negation. In conclusion, what ultimately enables us to live with others and for others is a crucial imperative: that we fully embrace activity that negates both the self and society and simultaneously allows us to realize our true task in response to that activity.

On the recovery of humanity in technology, we can think in the same way. As with the case of touch screens, new technology often alienates people who belong to minority groups. In order to recover humanity in technology, we should use technology not to alienate oppressed people, but for the inclusion of alienated people. Technology must be utilized for designing a new society which can be accessed and used by all people regardless of age, size, ability, or disability. Who is technology for? It is clear that technology is for a neighbor oppressed by our society. Our society has many difficulties which oppress neighbors. We use technology for changing our society and resolving difficulties for oppressed people. We must make ourselves strain our ears to hear the voices of oppressed people and use technology for them, not for ourselves and our society.

We can now understand the reason why Nishida mentioned "religion" and "God." He never asserted that we should convert to a specific religion. He said that we should change ourselves, and not try to create things only by ourselves. We should

---

[22] The word "species" results from TANABE Hajime 田辺元's criticism and his "Logic of Species". Cf. Tanabe (1963a, b). On Nishida' response to Tanabe's criticism, cf. Itabashi (2008: 85–184).

create things by hearing voices from oppressed people. Nishida regarded it as a religious imperative that we stop relying on our own power and make the voice of the other our guide. For him, the word "religious" means endeavoring to negate ourselves and obeying the ought.

Encountering "thee" and hearing the "thy" voice are important for technology. However, they are difficult for us. Let us give a concrete example. In Tokyo, a symposium entitled "The Philosophy of Technology for Co-existence" was held on March 10–12, 2006. Its theme was Universal Design. At the end of the second day, commentator KAWAUCHI Yoshihiko 川内美彦, who is a disabled person and uses a wheelchair, pointed out an important problem. Although the symposium was held in Japan, all presentations were given in English without translators. Furthermore, although its theme was freedom from barriers, there were no sign language interpreters or captioning services for people with auditory disabilities. This symposium was far from barrier-free and the Universal Design which aims to guarantee accessibility for all (Murata 2006: 202–203). MURATA Jun'ichi, one of the symposium's planners, confessed honestly that none of the planners had noticed the problems before Kawauchi pointed them out (Murata 2006: 13). Although they spoke about Universal Design, their actions were far from Universal Design. This case shows that it is impossible to change society for the better without hearing the voices of oppressed people. Even when we speak about Universal Design, we forget oppressed people. Thus, it is important for better technology to encounter oppressed people as "thee" and hear "thy" voice.

5 Conclusion

This chapter clarifies that to encounter "thee" is important for the recovery of humanity in technology. This "thou" means not a homologous "you," but the oppressed others. In modern times, going in the direction of subjectivism and anthropocentrism is inevitable. As a result, technology has lost its objectivity and become technology for its own sake. To take back objectivity, we must obey objective "ought." The content of the ought is to encounter "thee." For encountering "thee," we must realize that the historical schema continuing from the past that defines us is violently pushing "thee" into the position of enemies and outsiders. Thus, we must discover the true task in our society and break free from society. We must use technology for the "thee" who is oppressed by our society.

In Sect. 2, I discussed Nishida's characterization of humans as creative beings. Our creative action is "action-oriented intuition." This means that seeing is acting and acting is seeing. When we create something, we must look at nature. What unites ourselves and nature is technique. In Sect. 3, I discussed the problems of human reason. Because our reason can be beyond reality, our reason removes reality. As a result, our reason causes subjectivism and anthropocentrism. We lose objectivity and humanity. In Sect. 4, I elucidated how we can recover objectivity and humanity. They are recovered through the ought as objective expression. We cannot change the ought and we become ourselves by obeying it. This requires us to negate ourselves and our society to encounter "thee."

Of course, not all the problems of technology can be solved in this way; however, it is nonetheless necessary for solving any such problems.

References

Carew, Sinead. 2009. "Touch-screen gadgets alienate blind." *Reuters*, January 8. http://www.reuters.com/article/us-ces-blind-idUSTRE5080T320090109. Accessed 14 July 2017.

Carter, Robert E. 1997. *The Nothingness beyond God: An Introduction to the Philosophy of Nishida Kitarō*, second edition. St. Paul, MN: Paragon House.

Haver, William. 2012. *Ontology of Production: Three Essays*. Durham, NC: Duke University Press.

Heisig, James W., Kasulis, Thomas P., and Maraldo, John C., eds. 2011. *Japanese Philosophy: A Sourcebook*. Honolulu: University of Hawai'i Press.

House of Representatives 衆議院事務局, 2004. *Proceedings of the Committee Meetings on Financial Affairs at the House of Representatives in the 159th ordinary session of the Diet*第百五十九回衆議院財務金融委員会議事録, vol. 19. The House of Representatives 衆議院事務局.

Itabashi, Yūjin 板橋勇仁, 2008. *Historical Reality and Nishida Philosophy* 歴史的現実と西田哲学. Tokyo 東京: Hosei University Press 法政大学出版局.

Kosaka, Kunitsugu 小坂国継, 1995. *Nishida Kitarō: His Thought and the Present Day* 西田幾多郎:その思想と現代. Kyoto 京都: Minervashobō ミネルヴァ書房.

Krummel, W. M. John. 2015. *Nishida Kitarō's Chiasmatic Chorology: Place of Dialectic, Dialectic of Place*. Bloomington and Indianapolis: Indiana University Press.

Krummel, W. M. John and Nagatomo, Shigenori. 2012. *Place & Dialectic*. NY: Oxford University Press.

Murata, Jun'ichi 村田純一, ed., 2006. *The Philosophy of Technology for Co-existence: The Idea of "Universal Design"* 共生のための技術哲学:「ユニバーサルデザイン」という思想. Tokyo 東京: Miraisha 未來社.

Murata, Jun'ichi 村田純一, 2014. "The Creativity of Technology—Nishida Kitarō and the Philosophy of Technology技術の創造性:西田幾多郎と技術の哲学." In *Annual Book of the Nishida Philosophy Association* 西田哲学会年報, vol. 11. Nishida Philosophy Association 西田哲学会.

Nishida, Kitarō 西田幾多郎, 1965–66. *Complete Works of Nishida Kitarō* 西田幾多郎全集, edited by Shimomura Toratarō 下村寅太郎 et al. Tokyo 東京: Iwanami Shoten 岩波書店.

Ōhashi, Ryōsuke 大橋良介, 1995. *The World of Nishida Philosophy* 西田哲学の世界. Tokyo 東京: Chikumashobō 筑摩書房.

Okada, Katsuaki 岡田勝明, 2001. "'self-awareness' as 'Image': Problem between Nishida Philosophy and Imagination 「映像」としての「自覚」:西田哲学と構想力の問題." In *Japanese Philosophy* 日本の哲学 2: 44–61. Kyoto 京都: Shōwadō 昭和堂.

Shirai, Masato 白井雅人, 2006. "Self-awareness in the Historical World: On the Later Nishida's Philosophy 歴史的世界における自覚:後期西田哲学を巡って." In *Existence and Politics: Book Series of Existence Theory* 実存と政治-実存思想論集 21: 153–170. Tokyo 東京: Risōsha 理想社.

Shirai, Masato 白井雅人, 2007. "Negation and Sollen: Towards Developments in the Later Nishida 否定性と当為:後期西田哲学の展開に向けて." *Journal of Nishida Philosophy* 西田哲学会年報 4: 141–156.

Shirai, Masato 白井雅人, 2008. "The Locus of Logic in the Later Nishida: With the Body and Self-awareness as Leads 後期西田哲学における論理の場所:身体と自覚を手引きにして." *Philosophy* 哲学 59: 195–210.

Shirai, Masato 白井雅人, 2013. "The Fact of Self-Awareness and Its Development: The Question of Self-Awareness in the Later Nishida 自覚の事実とその展開—後期西田哲学における自覚の問題." *Journal of International Philosophy* 国際哲学研究 2: 105–112.

Shirai, Masato 白井雅人, 2014. "Religion in a Godless Age: The Question of Religion in the Later Nishida 神なき時代の宗教:後期西田哲学における宗教の問題." *Journal of International Philosophy* 国際哲学研究 3: 133–141.

Shirai, Masato 白井雅人, 2015. "A Forgiving God and a Judgmental God: The Question of Religion in the Later Nishida 赦す神と裁く神:後期西田哲学の宗教論." *Journal of International Philosophy* 国際哲学研究 4: 107–115.

Sugimoto, Kōichi 杉本耕一, 2013. *Nishida Philosophy and the Historical World: To the Question of Religion* 西田哲学と歴史的世界:宗教の問いへ. Kyoto 京都: Kyoto University Press京都大学学術出版会.

Suzuki, Daisetsu 鈴木大拙, 1989. "The True Man with No Title 無位の真人." In *The World of Suzuki Daisetsu* 鈴木大拙の世界, edited by HORIO Tsutomu 堀尾孟: 161–173. Kyoto 京都: Tōeisha 燈影舎.

Taguchi, Shigeru 田口茂, 2006. "Nishida Kitarō's Theory of the Free Will 西田幾多郎の自由意志論:自由と悪の問題をめぐって." *Journal of Nishida Philosophy* 西田哲学会年報 3: 87–102.

Takizawa, Katsumi 滝沢克己, 1972. *Fundamental Problems of Nishida Philosophy* 西田哲学の根本問題. In *The Works of Takizawa Katsumi* 滝沢克己著作集, vol. 1, edited by IIJIMA Munetaka 飯島宗享 et al.: 1–476. Kyoto: Hōzōkan 法蔵館.

Tanabe, Hajime 田辺元, 1963a. "The Logic of Species and the Schema of World 種の論理と世界図式." In *The Complete Works of Tanabe Hajime* 田辺元全集, vol. 6, edited by NISHITANI Keiji 西谷啓治 et al.: 1–31. Tokyo 東京: Chikumashobō 筑摩書房.

Tanabe, Hajime 田辺元, 1963b. "To Make Clear the Meaning of the Logic of Species 種の論理の意味を明にす." In *The Complete Works of Tanabe Hajime* 田辺元全集, vol. 6, edited by NISHITANI Keiji 西谷啓治 et al.: 447–522. Tokyo 東京: Chikumashobō 筑摩書房.

Chapter 3
A Moral Ground for Technology: Heidegger, Postphenomenology, and Watsuji

INUTSUKA Yū

1 Introduction

A man wakes up in the morning. He slips out of bed, pulls the duvet back over his sleeping wife, and goes into the living room. On his way to the kitchen to prepare breakfast for the family, he opens the curtains and blinds to let the light in. Eventually his wife and two teenage children, a son and a daughter, arrive and settle around the table. No one speaks during breakfast because their eyes are glued to their smartphones. The man becomes irritated and tells them to stop staring at their phones and enjoy their meal together as a family. His wife and children respond with little more than a distracted grunt.

Scenes like this take place every day all over the world. When a businesswoman receives an email related to her work, she is subconsciously preconditioned to respond immediately. The internet not only bombards us with news but pressures us to keep abreast or risk falling behind. Children of all ages are always checking their social media for notifications and updates on the assumption that preserving their friendships depend upon it. Today, emerging technologies surround us on all sides and there is nothing we can do to escape their demands.

A closer look shows that even some of our most ordinary ethical behavior depends on technology. In our opening example, when the man covers his wife with a blanket, opens the curtains to let in the light, and prepares breakfast for everyone, he is using the things around him (a duvet, curtains, light from outside, a table, dishes, bedroom, living

---

The original version of this chapter was revised: MS Mincho font was updated throughout the book for Japanese and Chinese characters. The correction to this book is available at https://doi.org/10.1007/978-3-319-59027-1_13

INUTSUKA Yū (✉)
Graduate School of Interdisciplinary Information Studies, University of Tokyo, Bunkyo, Tokyo, Japan

Thomas Taro LENNERFORS, MURATA Kiyoshi (eds.), *Tetsugaku Companion to Japanese Ethics and Technology*, Tetsugaku Companions to Japanese Philosophy 1,
https://doi.org/10.1007/978-3-319-59027-1_3

room, kitchen) to do what he thinks is good. He is using tools to manage his environment, but so are his wife and children as they tap away on their smartphones. On the face of it, there is no way to determine which acts are more ethical than others.

In recent years, discussions of the ethical aspects of technology have opened up a wide range of new questions and a variety of responses to them. For some philosophers, this has meant a growing pessimism regarding the harm that new technologies continue to inflict on society (Jaspers 1933; Ellul 1954; Marcuse 1964; Heidegger 1977). Others are critical of such pessimism, insisting that we must find a way to accept the role of technology and use it responsively (Feenberg 1999; Latour 2002; Verbeek 2011). Peter-Paul VERBEEK, a Dutch philosopher and the chair of the Society for Philosophy and Technology from 2013 to 2015, is a good example. Convinced that the autonomous human individual that Kant championed as the agent of ethical action is no longer viable, Verbeek has focused his research on the ethics of technology from an approach that Don IHDE has called postphenomenology. The foundational concepts of this approach break with those elements of phenomenology that distinguish it from the natural sciences in order to concentrate on describing the way the world appears to us (Verbeek 2011: 14). Instead, they are more concerned with the relationship between human beings and the world, and on the role that technology plays in that relationship. Verbeek's 2011 book, *Moralizing Technology*, marked a turning point in this development, drawing on thinkers such as Don IHDE, Bruno LATOUR, Michel FOUCAULT, and others.

Against this background, in this essay I would like to raise the question of good and evil. While Verbeek criticizes the idea of an autonomous ethical subject independent of the growth of technology, he implies that living the good life relies on the decision of the conscious individual. But can individuals, or a collection of individuals, really determine what is good or bad if the autonomy of the subject is in doubt?

I would like to propose the ethical philosophy of the modern Japanese thinker, WATSUJI Tetsurō 和辻哲郎 (1889–1960), as one way to respond this question. Like Verbeek and so many others, Watsuji was strongly influenced by Heidegger's analysis of equipment (*Zeug*) and his subsequent rejection of the autonomy of the individual person. For Watsuji the moral agency is a complex dynamic involving both humans and their environment, and it is there that good and evil are seen to emerge. As we shall see, the notion of trust is particularly significant here.

From there I will go on to examine Heidegger's analysis of equipment and compare it to Verbeek's ethics of technology. In this context, I will outline Watsuji's take on Heidegger and consider its consequences for the ethics of technology. I will conclude with a comparison of Verbeek's and Watsuji's claims as they relate to the ethics and technology today.

2 Heidegger and Verbeek

Are tools neutral? Take the case of a gun. In the hands of someone who uses it to shoot someone else, is it only a tool? Does all the responsibility lie with the one who uses it?

Heidegger's analysis of "equipment" in *Being and Time* is an important resource for philosophers dealing with technology. As he sees it, the true nature of human existence as Dasein does not require an opposition between the human subject and the external world. On the contrary, Dasein is itself a "being-in-the-world." What is human about our existence actually resides "outside" (Heidegger 2006: 62).[1] We live amidst the things in the world. Of our various ways of "dealing" with things, it is not perceptual cognition but a "concern that manipulates things and puts them to use" that best marks our human encounter with "equipment" (Heidegger 2006: 67–68). Equipment does not refer to discrete objects but to a complex of interrelated objects like pen, paper, and desk (Heidegger 2006: 68). Together they serve as "something in-order-to." The use of equipment entails an "assignment" of one thing to another. For example, when I use a hammer, a hammer becomes a part of me, but it is not my main concern. By using it my attention is assigned to a finished product, say a pair of shoes (Heidegger 2006: 69). The finished product, in turn, represents another piece of equipment in the sense that shoes are made to be worn. The product I have in mind to make assigns the "materials"–in our case, leather and nails. Leather is made from the hides of animals. Hammer and nails are made of metals, which in turn are refined from minerals in the earth. In this sense, nature is disclosed in the use of equipment (Heidegger 2006: 70). Finally, the end product assigns others as its users, namely, people who wear shoes. My contact with the equipment does not itself generate this nexus of assignments, because some of the equipment may be of no use or even missing (Heidegger 2006: 73). Even its absence confirms the fact that we are inseparable from the world.

Ihde (1990) adapts Heidegger's analysis of the relationship between Dasein and equipment to analyze how "technological intentionality" shapes perceptions and experiences (141). He suggests four main ways in which technologies mediate our interaction with the world (Ihde 1990: 72–112). First is the "embodiment relation," wherein tools – everything from eyeglasses to automobiles to Heidegger's hammer – actually alter the way we perceive the world. The second, "hermeneutic relation" points to the way we interpret the world by reading things like books and thermometers. In the third, "alterity relation" technological media confront us as autonomous entities, as in the case of a personal computers. The fourth and final way in which technologies mediate the world to us is the "background relation." Things like housing, lighting, heating, and clothing are all tools that work in the background of daily life.

Verbeek then goes on to introduce the question of ethics, beginning with a critique of pessimistic approaches that pit technology against the quality of human life. Ihde, Bruno Latour, and others had already questioned the assumption that humanity exists independently of technology. Their work demonstrated how technology so deeply permeates human activity that it can never be seen as completely neutral. If there were no guns, no one would think about shooting anyone else. Verbeek (2011) drew the conclusion that morality is embedded within technology itself. He calls this "material morality" (viii). When Latour pointed out in 1994 that speed bumps

[1] For English translations of Heideggerian terminology used in this section, see Heidegger (1962).

replace the need for individual morality by forcing drivers to slow down, he was giving an example of just this. Verbeek himself uses the example of an ultrasound scan, which establishes a connection between the parents and their unborn child. At the same time as it enables an emotional bond, it may also trigger the choice to abort the fetus for medical reasons. Without taking into account, the mediating role of technology, there is a range of ethical problems we cannot properly address.

Verbeek then goes on to adapt and expand an ideas drawn from Michel FOUCAULT. For Foucault, humans depend on power and it is through active engagement with this power that we foster the good life. Given that we depend on technology, Verbeek argues that living a good life requires us to engage with it actively. This differs from modern Western ethics with its fixation on right action by reestablishing contact with classical and medieval virtue ethics. In doing so, it seems to me, he goes well beyond a mere "ethics of technology" to propose a major shift in the field of ethics, as when he writes;

> A few centuries ago the Enlightenment, with Kant as its major representative, brought about a turnover hitherto unequaled in ethics by moving the source of morality from God to humans. Do contemporary analyses of the social and cultural role of technology now urge us to move the source of morality one place further along – considering morality not a solely human affair but also a matter of *things*? (Verbeek 2011: 12)

Verbeek raises an important question in suggesting we broaden the "source of morality" to include both the human and the non-human. At the same time, he admits that "in our liberal democracy, the freedom of the individual is a very important value, but the quest for answers to the question of the good life belongs to the private sphere rather than a public space" (Verbeek 2011: 112). Because he gives technology a mediating role, he insists on shifting the realm of good and evil from the private sphere to the public. "Technologies bring the question of the good life back into the public sphere – for the simple reason that they embody and help to shape visions of the good life" (Verbeek 2011: 159). Against a "technocracy" in which an engineer constructs an ethical framework for a society, Verbeek advocates an extended version of Constructive Technological Assessment (CTA) to balance diverse values (Verbeek 2011: 102). CTA is a method used to collect evaluations of technologies under design by bringing together a variety of agents including users, pressure groups, engineers, and firms to reflect on the results as they relate to the design. Verbeek calls this the "democratization of the design process" (Verbeek 2011: 103). He further suggests a new task for CTA: to "anticipate possible mediating roles of the technology-in-design" (Verbeek 2011: 103). This, in turn, requires developing a public space and to discussing questions about the good life (Verbeek 2011: 159).

> Without seeking to return to a situation in which the question of the good life is answered by a state or a church, the analysis of moral mediation developed in this book urges us to acknowledge that technologies bring the question of the good life back into the public sphere – for the simple reason that they embody and help to shape visions of the good life. (Verbeek 2011: 159)

An example of the kind of public discussion Verbeek has in mind is the way in which technology affects the unborn through things like obstetric ultrasounds. His efforts are an important step in offsetting the emphasis on individual freedom by encouraging a wider exchange of ideas regarding good and evil.

This is where Watsuji come into the picture. Verbeek has expanded moral agency from the individual to a complex of human beings and their technological artifacts. He has also shown how new technologies engender new human relationships and hence also new questions of good and evil. But despite bringing the non-human into ethics, he seems to retain the role of individual consciousness in distinguishing between good and evil. Even his expansion of the role of CTA and public discussion seem to be grounded in individual judgment. When all is said and done, the various notions of the good belong to the persons that espouse them. In other words, he does not sufficiently explain how nonautonomous individuals can judge an action or a life to be "good."

When non-human technologies are included in moral arbitration, the absolute authority of the human element necessarily comes into question. Individuals, alone or collectively, can no longer act arbitrarily. At the same time, there are no moral universals transcending time and place. The same can be said of particular complexes of the human and the non-human. If judgments of good and evil are themselves phenomena appearing in such a dynamic complex, other ethical questions arise. Watsuji brings all of this into clearer relief.

3 Watsuji's Critique of Heidegger

At several points in his writings Watsuji challenges Heidegger. His criticisms vary in frequency and intensity, but on one point he is consistent: Heidegger's analysis of equipment pays insufficient attention to the developmental history of technology. For Watsuji this history is further proof of how deeply human life is rooted in the natural environment and the social character of human existence.

Watsuji's critique can be traced back to a set of lecture notes that bear the title "Thoughts on National Character." The lecture itself seems to have been presented at Kyoto Imperial University between 1928 and 1929, shortly after his return from Germany. In it Watsuji focuses on the geographical differences between human societies, in contrast with the Marxist account, which focuses on their historical development. In his defense, he draws on Heidegger's analysis of equipment through consideration of the relationship between human beings and their surroundings. Watsuji then offers four criticisms.

First, he questions Heidegger's idea that the most immediate form of "dealing" is a "concern that manipulates things and puts them to use" (WTZ-B1: 388). Heidegger took the example of a hammer to show how Dasein mediated by equipment uses the physical environment for a specific purpose. But, Watsuji suggests that if we take the example of a house or clothing, it is hard to see how "concern to use" is primary. The experience of things like "cold" and "rain" precedes it. Here

Dasein is not manipulating things or otherwise using them, but rather acting in response to what is encountered in the world.

A second and related criticism of Heidegger's analysis is that everything encountered in the world is viewed in terms of equipment (WTZ-B1: 389). Cold and rain may have their role to play in the production of foods, but wind is first and foremost experienced as a cold wind or a soft and gentle breeze, not as a means to billow a sail or turn a windmill.

In the third place, Watsuji questions the validity of identifying "concern" (*Besorgen*) as a way of being-in-the-world particular to Dasein (WTZ-B1: 390). "Enjoyment" also needs to be included as part of our dealings with the world. We choose clothes made of wool or soft cotton because we like the way they feel and not just because they keep us warm. One might classify this as equipment designed "to satisfy the sense of touch," but the interaction involved is not so much one of "concern" but as "enjoyment." The same may be said of all equipment. If we think of enjoyment or sensation as a way of dealing with the world, we need to take into account both the active and passive aspects of the dealing. We can find what we seek and we can miss what we don't find, but we can also find what we were not looking for at all. All these dealings bring their own understanding, but the understanding enjoyment or sensation is more fundamental than the understanding of simple use. Before you understand what a sweater is, you must understand what cold is.

Finally, Watsuji takes up the idea of mood (*Stimmung*), which Heidegger considers an everyday "disposition" and in which he locates the *da* of Dasein (WTZ-B1: 392). For Heidegger, mood seems to appear out of nowhere, but Watsuji is not convinced. Clearly our moods are directly related to how we experience our natural surroundings (*Umweltnatur*). As he sees it, it is not that fresh morning air by itself makes us feel refreshed. Both the air and our Dasein have a part to play in creating the mood of feeling refreshed.

For Watsuji, tools and the self-understanding of Dasein are both grounded in sensations associated with our natural surroundings. He developed this approach in a 1929 essay entitled "Climate [*Fūdo*]" and later expanded into a book of the same title.[2] In it he writes:

> It goes without saying that, as equipment, clothes, food, and shelter take on a climatic character; but even more fundamentally, if human beings are already subject to the regulations of climate at the time they discover themselves, then the form that climate takes must finally shape the *form of self-understanding* as well. (WTZ 8: 21–22; Watsuji 1988: 15, translation adjusted)

Here Watsuji goes a step further by suggesting that being human is not an individual ideal but a collective one. We do not ex-sist in the cold as individuals but as a community of human beings. The very act of protecting ourselves or others against the cold, as when we put a coat on a child, implies an understanding that we are in the cold together. Such shared understanding accumulates through climatic repetition and expresses itself in the way we cloth and house ourselves. Watsuji refers to

---

[2] *Fūdo: Ningengakuteki kōsatsu* 風土 人間学的考察 (*Climate: Thoughts on the Study of Human Being*). 1935.

this distinctively human bond to climate as "climaticity" (*fūdosei* 風土性) in contrast to Heidegger's concept of "historicality." Watsuji is reacting to the individualistic tendency of Heidegger's Dasein, which glosses over the spatial aspects of human existence in favor of the temporal.

After returning from Germany, Watsuji concentrated his attention for some years on ethics. In 1931 he published a lecture entitled "Ethics."[3] Three years later *Ethics as the Study of Human Being*[4] appeared, followed by his most important work on the subject, the three-volume study called simply *Ethics*[5] (1939–1947). These texts focus on the social nature of human beings, which in turn provides a standpoint from which to criticize the individualizing tendency of Heidegger's Dasein. "To live in the world means, first of all, to live in a communality of human beings as a society, and only later involves dealing with the world of things as equipment" (Watsuji 1931: 60). Human interaction is prior to interaction with things. Watsuji plays on the Japanese word for furniture, *kagu* 家具, literally house implements, to observe that there must first be a *kazoku* 家族, or household, for there to be things to belong to. Returning to the example of the hammer, Watsuji sees it as an implement belonging to society (WTZ 9: 162).

> Somethings like dealing with tools is only possible because the tools themselves are already *grounded in human relationships*… Even if we take the example Heidegger often uses of dealing with handicraft tools, these dealings only become real when grounded in the organization of handicraft labor in the sense of the human relationships involved in activity… He further argues that it is only through the medium of these tools and their products (which are another kind of tool) that other persons become manifest. What he does not see is that the tools themselves emerge *from human relationships* involved in labor. Even if we grant that the steps in Heidegger's account are in line with ontology, the structure of "being-in-the-world" is not enough to justify taking dealing with tools as a point of departure rather than begin from dealings with other persons. We find tools only in our dealings with others. (WTZ 10: 185)[6]

In his treatment of equipment, Heidegger also describes others as a "Dasein-with" (*Mitdasein*), but they are little more than a collection of atomic individuals. For example, to the cobbler, others would simply be individuals who wear shoes. Thus, Heidegger's ontology "misses the practical action-nexus, namely, the human existence that reaches deeper than the understanding of being" (WTZ 9: 162). "Practical action-nexus" is a term peculiar to Watsuji's ethics. It refers to the operational nexus of activities as the dynamic complex of so-called human beings and their physical environment, including technologies.[7] Together they constitute "human existence" as a whole, whose relationship to the good life we turn to next.

---

[3] "*Rinrigaku* 倫理学" in *Iwanami kōza, Tetsugaku* 岩波講座 哲学 (*Iwanami Lecture Series of Philosophy*). 1931.

[4] *Ningen no gaku toshiteno rinrigaku* 人間の学としての倫理学 (*Ethics as the Study on Human Being*).

[5] *Rinrigaku* 倫理学 (*Ethics*).

[6] The English translation of this passage (Watsuji 1996: 175–76) seems to have missed Watsuji's meaning on several points.

[7] Yamamoto and Carter translate *jissenteki kōiteki renkan* 実践的行為的連関 as "the practical interconnection of acts" (Watsuji 1996: 36).

4 Watsuji's Ethics

Watsuji defines ethics as the "study of human being," which includes reexamining human existence in order to confront the problem of good and evil. The whole system takes up three massive volumes, but here we are interested mainly in two propositions: (1) an agent of action is an "action-nexus" or a dynamic complex of humans and their environment; and (2) good and evil have to be understood in terms of "trust."

4.1 Betweenness and Action-Nexus

One of the most important concepts for understanding Watsuji's *ningen sonzai* 人間存在, or human existence, is that of *aidagara* 間柄. The Japanese word *aidagara* denotes relationship between human beings. It is made up of *aida* – a space between things, a temporal interval, or a relationship – and *gara* – which is a pattern or essential character. Watsuji considers multiple meanings of *aida* help to illuminate the word Japanese uses for human being: *ningen* 人間. Literally, the two characters express person (人) and betweenness or relationship (間).

> The locus of ethical problems lies not in the consciousness of the isolated individual, but precisely in the *in-betweenness of person and person*. Because of this, ethics is the study of *ningen*. Unless we regard ethics as dealing with matters arising between person and person, we cannot authentically solve such problems as distinguishing between good and evil deeds, obligation, responsibility, virtue, and so forth. (WTZ 10: 12; Watsuji 1996: 10, translation adjusted)

Watsuji's aim here is to construct an ethics, or *rinrigaku* 倫理学, on the basis of two distinguishing traits of human beings, namely sociality and individuality (WTZ 10: 16). Human beings are both a "society (*yo no naka* 世の中)" and the "persons" that make up that society (WTZ 10: 17).[8] What I think Watsuji is proposing is that human existence is not a mere collection of individual bodies but something that opens up out of the embodied individual.

Watsuji's approach derives from an analysis of "action." Obviously relationships among humans differ from relationships among objective things. But humans are subjects of "action-nexus." It is not just that persons cannot be in relationship without acting subjectively; they cannot act at all except in a relationship. This is what it means to refer to a human relationship as a subjective action-nexus (WTZ 10: 20–21). For Watsuji, *aidagara* and action-nexus are synonymous (WTZ 9: 25).

The notion of "action" also highlights the difference between the way people interact each other and the way they interact with things. When we see something, it is we who do the seeing and the thing that is seen. But when we see one another

[8] Watsuji is probably contrasting the Japanese term *yo no naka* 世の中, which means literally "in the world" but generally means a society, to Heidegger's "being-in-the-world."

as humans, Watsuji notes, things get more complex. We can look at each other, we can stare or leer or ogle or even just steal a glance. All these ways of seeing entail some nexus of action which is necessarily mutual and never unilateral. Watsuji means to distinguish these properties of *aidagara* as action-nexus from ordinary phenomenological intentionality (WTZ 10: 36).

What determines the way people see each other? Watsuji's answer is to draw a distinction between totality and individuality. Totality is the specific context that determines an individual's actions (WTZ 10: 95). Take the father of a family who works in an office. We distinguish between the things he does for his family and the things he does for his company by understanding the context within which they take place. Or again, the action of cutting into a body is justified when it is the action of a surgeon performed on a patient, but not in the case of a criminal stabbing. In each case, the totality does not exist without the actual presence of its members. The relation of this totality to its members is not like the relation between a body and its limbs for the simple reason that the members may move in and out of the totality. When a parent does not behave as a parent, the totality of the family is lost (WTZ 10: 95). In short, neither the totality nor the individuals exist by themselves but only in their relationship to one another (WTZ 10: 106). Theirs is a relationship of negation: an individual has the character of an individual because of its ability to violate the totality, and a totality has the character of a totality because of its ability to negate the independence of the individual. To understand human interdependency, Watsuji claims is to grasp the fundamental emptiness (*kū* 空) that lies behind it (WTZ 10: 124).

Furthermore, Watsuji's notion of action-nexus is applied to time and space. A certain sense of spatiality comes into play when we consider a nexus of human subjects engaged in practical activities. The unity or opposition between two persons creates an inside and an outside that differs from physical space (WTZ 10: 164). This subjective spatiality is a fundamental trait of Watsuji's *ningen* and *aidagara*. In the public sphere, too, spatiality is not just physical. It is the locus of a practical nexus that can be modified or even distorted. This nexus includes everything from mass media to the flow of traffic (WTZ 10: 165). Subjective spatiality as such is a function of the two aspects of human existence we referred to earlier, namely, sociality and individuality (WTZ 10: 173). It is generated as human existence breaks off into multiple individuals which then reunite in one or the other form of nexus. To return to what was said earlier, for human relationships to serve as a basis for our relationship with things, we must also reconsider the meaning of space in terms of human relationships. In the final analysis, physical space itself is grounded in human spatiality.

Much the same can be said of time. If people exist spatially, they also exist temporally (WTZ 10: 189). Take traffic at its most basic level: people walking to and fro from one place to another, from one human relationship to another. The act of walking to a friend's house or to one's place of work already implies a human relationship with both a past and a future, a movement toward a new human relationship based on a former one. Within this dynamic complex of past and future, repetition occurs. This gives rise to the fixing of "times" or opportunities for certain activities. We have a time to wake up and a time to go to bed, a time to work and a time to eat, a time to sow and a time to harvest (WTZ 10: 200). For society to function, we need to coordinate these activities according to publicly accepted conventions of time. In former ages, the movements of the sun and the moon gave us the hours of the day, the days of the month, and the months of the year. The idea of a universal, objective time is based on archaic units of measurement, which, in turn, are grounded in human relationships (WTZ 10: 202).

Having examined Watsuji's perspective on the prevalence of human relationships in the world, including in our understanding of space and time, we now turn their role in technology. We begin by analyzing his notion of "expression" (*hyōgen* 表現). In addition to drawing on the reflections of Heidegger, Watsuji relied heavily on Dilthey's hermeneutics to explore the connections between ethics and technology. In his most important works *The Formation of the Historical World in the Human Sciences* (1910), Dilthey points to a fundamental problem in the human sciences. Unlike the natural sciences, which deal with external objects, the human sciences are essentially reflexive, which raises a problem: the subjects are also the objects, but human subjects can never really be objects for themselves. Dilthey resolves the paradox by introducing the idea of a "nexus of life, expression, and understanding" in order to argue that life and the lived experiences of human beings are objectified as "expressions" in the form of words, gestures, or deeds. He argues further that life is only accessible to us by projecting our own experiences onto such expressions. Watsuji recognizes the force of the same paradox in his own ethics. But while he accepts the basic structure of Dilthey's threefold nexus, Watsuji criticizes him for limiting his analysis to individual experiences and their modes of expression as found, for example, in works of art or biographical accounts (WTZ 10: 47). Watsuji suggests that *all* expression in human life – including our experience of the natural world – is an "expression of *aidagara*," (WTZ 10: 38) and that points to "practical action-nexus" (WTZ 10: 39) as the basis of our every experience.

Technology (*gijutsu* 技術) does not figure dominantly in Watsuji's texts, though he does introduce the idea of equipment or tools (*dōgu* 道具), and even "technologies" (in the plural) as expressions of *aidagara*. In this way, the architecture of a house can be seen as an expression of human existence within a family (WTZ 9: 163–64). In this context, doors without keys express *aidagara* differently from doors with keys. Or again, a dining room, table, and dishes express a part of family living that needs to be understood if its members are to sit down to a meal together:

> Equipment is already an expression of *aidagara* and not merely "something to hand" for me. Thus we seek channels to human existence in daily expressions of that existence such as looking at one another, talking to one another, interacting one another, as well as in the various ways these daily relationships are expressed in material objects. Our everydayness rests on an understanding of all these modes of expressions. (WTZ 9: 162)

Watsuji's idea that tools are expressions of *aidagara* is similar to Verbeek's idea of material morality (Verbeek 2011: viii). But whereas Verbeek restricts his attention to "technological artifacts," Watsuji broadens the field to include natural phenomena like cold, rain, and mountains. The *aidagara* present between a mother and her child encompasses "environmental expressions" (*kankyōtekina hyōgen* 環

境的な表現) like the woods and streams around their house that are a part of their past together (WTZ 11: 7). In the same way that tools mediate our actions as expressions of *aidagara*, so does the natural environment. The environment does not stand opposed to social structures; it belongs to their very existential content (WTZ 11: 152). The fields that a farmer cultivates for his family belong to the nexus of the family:

> A local landscape does not indicate an area where a community has left its mark on *the natural world that stands opposed to it* but a place where the community *uses the land to express the content of how it structures human relationships*. (WTZ 11: 154)

In this way Watsuji argues that human existence is to be understood as an action-nexus that unfolds within a complex framework of technologies interacting with the natural environment. "Understanding expressions of an *aidagara* develops the *aidagara* itself" (WTZ 9: 144). An action-nexus allows for countless forms of expression and understanding. By sharing these expressions among ourselves, we actualize the nexus of self and others (WTZ 9: 151).

4.2 Trust

As we have seen, Watsuji expands the scope of human existence to embrace space, time, and the physical environment, and he does so as part of his overall effort to counter the idea that human existence can be analyzed merely in terms of individuals or only in terms of societies. It is *both* individual and social, an action-nexus that unfolds dynamically through repetitive expression and understanding. This brings us back to the problem we began with: how can good and evil be determined if neither the individual nor the totality of a society qualifies as an independent moral agent but are both fundamentally "empty"?

To address this problem, we must first look more closely at Watsuji's analysis of "trust," which he locates in the context of the spatiality and temporality of human existence in action. When someone asks for help, for example, they already expect someone else to act (WTZ 10: 279). Or again, people can walk in a crowd without protection because they expect that others will not assault them. In much the same way, an office worker or a teacher carry out their daily activities on shared assumptions about their respective social position (WTZ 10: 281). None of this would be possible without a certain foundation of mutual trust. All of these examples are connected to spatial quality of the unity of self and others, but temporality also has a role to play in trust. We are able to sustain relationship with others into the future because the past that survive in the present also carries on into the future (WTZ 10: 285). Taken together, the spatial and temporal structures of action constitute what Watsuji calls a "movement of coming back" (*kirai no undō* 帰来の運動),[9] which is said to serve as the foundation of trust in human relationships:

[9] He seems to be referring indirectly to Heidegger's idea of the "coming-back" (*Zurückkommen*) of Dasein (Heidegger 2006: 326).

> The ground of trust lies in the spatial and temporal structure of human existence. In other words, it is the rationale (*rihō* 理法) of human existence unfolding spatially and temporally that makes trust possible. Consequently, the seemingly credible proposition that "human relationships are built on trust" actually has things upside down. The rationale behind the spatial and temporal nature of human existence is the basis of human relationships and, therefore, the *ground* of trust… Thus, human relationships are at the same time relationships of trust, and wherever there is a human relationship, trust arises at the same time. This is not to say that there is no such thing as distrust or of betrayal, but rather that they represent an absence of trust, and therefore they are negated at the deepest levels of the existence as an offence against the very rationale of human existence. (WTZ 10: 285–86; Watsuji 1996: 271–272, translation adjusted)

From of old, betrayal has been rejected as sinful. But what happens to bring about a loss of trust? To answer this question, Watsuji turns to a discussion of "truth and falsehood." The truth of human existence is no more a matter of the individual, no more a matter of transspatial or transtemporal judgement than the rest of human relationships (WTZ 10: 286). In the end, truth and falsehood are circumscribed by human relationships. If truth is often taken as a correspondence between words and facts, experience suggests that things are not quite so simple. Shielding a sick friend from the reality of their illness may be done out of consideration for the person and not necessarily out of the desire to deceive (WTZ 10: 289). Similarly, when it comes to the correspondence between one's words and one's actions, the kinds of promises people make to each other with regard to the future affect truth and falsehood (WTZ 10: 289). Simply put, truth arises out of a relationship of trust. Falsehood cannot coincide with the truth but neither can it exist apart from it. Falsehood signals the absence of truth, or more precisely, the pretense of truth (WTZ 10: 295–96). When it takes the form of fraud or deception, falsehood disables the bond of unity between persons, and at least for the moment blocks the process of "coming back" to fundamental authenticity of *aidagara*. This also holds at the level of totality. Relationships of trust cannot come about in a society where individuals have no free choice but are merely appendages subject to a greater organic whole (WTZ 10: 297). Trust requires the tension of a certain independence from the totality. In a word, the distinction between truth and falsehood reinforces the primacy that Watsuji accords human relationships.

In this context Watsuji takes up the question of good and evil with the simple proposal that "responding to trust and enabling the truth is good, while preventing the truth and enabling falsehood is evil" (WTZ 10: 302; Watsuji 1996: 286). Standards of good and evil may be thought to vary according to time and place, but the more basic differences lie in how far relationships of trust extend and how they are expressed. The principle that good is a response to trust and evil its betrayal remains constant (WTZ 10: 308). That said, the apparent evil of an act of betrayal may turn out, in time, to have been for the best (WTZ 10: 300). Watsuji understands "radical evil" (*kompon aku* 根本悪) as absolutizing the individual or the totality, and thus putting a halt to the process of the "coming back" to authenticity of *aidagara*. Human existence is twofold: it is individual and it is social. If either overwhelms the other, human existence belies its fundamental nature (WTZ 10: 298).

Watsuji associates guilt with the betrayal of trust among human beings. As he understands Western thought, ideas of guilt once associated with a betrayal of God and the laws of God have given way in modern philosophy to more individualistic and radically privatized ideas of guilt (WTZ 10: 314). Heidegger's views on guilt as the origin of morality are cited as an example. For Heidegger, the finite and groundless nature of Dasein engenders a sense of lack or emptiness in existence, which lurks behind feelings of indebtedness or guilt (*Schuld*). This sense of guilt sets up an ontological constraint out of which morality arises. For Watsuji, this progression of ideas seems entirely arbitrary (WTZ 10: 315):

> There is no debt if I am not indebted to anyone for anything. In the same way, I cannot be indebted to others if I myself am nothing. What is more, if this empty self is accompanied by a sense of guilt, it is not because the self is perceived as an empty nothing but because of feelings of hatred, distrust, disobedience, and the like, which are experienced as *betrayal*. (WTZ 10: 317; Watsuji 1996: 298)

Watsuji appreciates Heidegger for connecting guilt and conscience to the temporality of human existence (WTZ 10: 326), but he argues that the failure to integrate temporality and spatiality inhibits him from a thoroughgoing treatment of conscience. The voice of conscience speaks to us when we fail to do something we know we ought to do. Conscience always has to do with action. Conscience is a call from our fundamental authenticity; it appears when we betray trust and eclipse the truth that can only be found in a return to that authenticity (WTZ 10: 325). Thus the disclosure of the authentic self that Heidegger attributes to conscience is actually the disclosure of a totality already present within the self (WTZ 10: 326). This is what Watsuji means when he says that conscience is rooted in both the temporal and spatial structure of human existence (WTZ 10: 328).

5 Verbeek and Watsuji

As we have seen, Verbeek and Watsuji have cast a critical eye on Western ethics and its traditional concentration on the conduct of the individual moral agent. Both of them exclude the idea of the autonomous individual from their ethics on the grounds that moral agency does not begin and end with the skinbound human entity but belongs to a greater complex that includes both the human and the non-human. The influence of Heidegger is evident in their philosophies, as are developments in hermeneutics after phenomenology.

There are two points, however, on which Verbeek and Watsuji differ fundamentally: what constitutes the non-human and where good and evil come into the picture. For Verbeek, moral agency can be extended to include technological artifacts as well as human beings. Watsuji goes a step further to give the natural environment a mediating role in human morality. On his view, both technological artifacts and the natural environment are seen as critical elements of human existence or action-nexus; they are an "expression" of *aidagara*. While questioning the autonomy of the individual, Verbeek nonetheless assumes that conscious subjects are able to discuss the good life and make decisions about it. Watsuji, in contrast, insists that the individual is essentially incapable of determining what is good or evil, since good and evil only become apparent in relationships of trust: the good is a positive response to trust, and evil, a negative response. Because trust is based on past relationships and the process of "coming back" to them, individuals detached from the past have nothing on which to base a relationship of trust.

Given the characteristic of technology to mediate human activity, Verbeek proposes CTA as a way to negotiate different views of the good life. At the same time, he encourages people to come together on common ground to discuss their views and share ideas. His underlining assumption seems to be that people can make up their own minds about what constitutes the good life. But can an idea of the good formulated by individuals truly be said to represent the good?

Watsuji's response is a clear no. Behind every view of the good life lies a relationship of trust shaped by past generations. Following this line of thought, we might say that the reason there are so many views on the subject nowadays is not that individuals are more willing to express their private opinions, but that a plurality of pasts have been brought into closer contact with one another. The kinds of fundamental ethical questions Watsuji is asking can also be applied to new technologies in our own day. We all know what a profound impact technology has had on the rhythm of our lives. The past is in danger of being edged more and more out of the picture and plurality of being replaced by uniformity. In Watsuji's terms, we might call this an interruption in the process of "coming back" to the foundations of trust on which good and evil are based. His admonitions with regard to technology apply to large scale structural systems and equipment as well as to everyday tools like dishwashers and smartphones. Technology has the potential to transform our sense of space and time drastically, and thus to rupture existing human relationships and compromise the trust that holds us together. Watsuji recognizes that the action-nexus for a given individual can be exceedingly complex and involve relationships of trust across a number of contexts, with the result that trust and betrayal can exist side by side. A politician may work diligently to honor the trust of his constituents, all the while carrying on an extramarital affair that betrays the trust his family has placed in him (WTZ 10: 297). Advance in technology often create and sustain these complications within everyday morality.

For Watsuji, moral agency is rooted in a shared apprehension of the natural environment. But as modern technology isolates us more and more from our surroundings, it endangers that original connection. Even the vegetables on our dinner table are a sign of our ongoing alienation from the natural world. From the genetic modification of the seeds to the chemical fertilization of the fields to the global agricultural conglomerates that cultivate the crops and transport them around the world, technology has in effect walled us off from the vegetable garden. Try as we might to live the good life, the basic conditions for living take that ideal away from us. The totality of human life is fragmented, and without the environment that once served as the commons for human relationships, we are left to our own individual devices. This is what Watsuji calls radical evil. Verbeek reproaches ethicists who take a pessimistic view of technological progress, but insofar as that progress marks a break in the bonds between the human community and the natural world around them, a certain level of suspicion is not only reasonable, it is necessary. As our shared past becomes more and more unreliable, the future of our shared humanity grows more and more unpredictable.

Whatever the future has in stored one thing is certain: ethical action will be mediated by the tools we have at hand. For this reason alone, Verbeek is right to urge us to engagement with technology. For Watsuji any such engagement entails sustaining the rhythms of life that hold us together and enable trust. Verbeek's idea of a CTA needs to expand to include this aspect, as well as to take into account the way in which the rhythms of the natural world serve to channel the various rhythms of a society and the individuals that make it up. Surely he recognizes, as did Watsuji, that human beings have always given order to their lives by coordinating its rhythms with those of the natural environment (WTZ 11: 16–17). At the same time, we cannot ignore the fact that the expansion of consumer capitalism combined with the advance of technology has led to what Jonathan Crary (2013) calls "the end of sleep." The effect on the relationship of trust is, not surprisingly, disastrous.

Watsuji's efforts to deprivatize ethics merit further investigation. In addition to the ethics of technology, which has been the center of our attention here, discussions today of environmental ethics may find Watsuji's thought suggestive of ways to resolve the standoff between the human and natural worlds. The compelling feature of Watsuji's ethics, it seems to me, lies in its integration of the mediating role of technology and the natural environment in human moral agency.

6 Conclusion

Today, perhaps more than ever before in history, technology has become inseparable from daily life. Ethical theories up until now have tended to set human beings and technology in opposition to one another and to warn us of the dangers that tools pose to the quality of life. Recent trends in ethics have taken a more positive approach to technology as a mediator of morality. Verbeek is representative of this trend. Drawing on Ihde's research with its grounding in Heidegger's philosophy and on thinkers like Latour who stress interaction of the human and the non-human, Verbeek's efforts to engage ethics more closely with technology are an important part of the picture. In this connection, he introduces the idea of Constructive Technological Assessment in order to shift the question of the good life from the private sphere to the public. It is not that he wants to do away with the idea that decisions about what is good, and why, belong to individual subjects. But the question of good and evil does not rest finally on the autonomy of the individual. Somehow, the connections between the human and the non-human need to figure in the discussion. This was my starting point in this essay.

I began with attempting to relate the central concerns of Watsuji's ethics to the complexities of the kinds of question Verbeek and others have been asking. In his critiques of Heidegger's analysis of equipment, Watsuji proposes the natural environment and social relationships as a basis for treating technology and ethics. Our twofold nature as individual and social affects the way we encounter the world. For Watsuji, therefore, we can understand our natural surroundings and the tools we use as expressions of *aidagara*. Human existence is a dynamic action-nexus that develops through a return to *aidagara* in a movement from the past to the future. This is the basis of trust, which in turn determines good and evil as a response to or betrayal of trust.

If we accept Watsuji's logic, the question of technology can be rephrased as a problem of trust. Insofar as technological innovation disrupts the rhythms of daily life, it undermines relationships of trust. I began my remarks with the image of a family in which the husband, the wife, and the children were all trying, each in their own way, to live up to the trust of others – the wife with her colleagues and the children with their friends. The problem was that this conflicted with their father's expectations for the family. In a sense, the sort of multiple personality disorder depicted in this example is a reflection of a general social reality which advances in technology only seem to aggravate. Verbeek's insistence that leading a good life requires us to engage with technology speaks directly to our situation. Watsuji has led us to see that the good is not entirely a matter of individual choice for an individual life, but involves maintaining a relationship with the shared past as a basis for trust. A productive and promising future for us as individuals and as a society begins with awareness of how severely fragmented our lives have become and how far we have strayed from the *aidagara* that is our primordial gift as human beings.

Acknowledgement This work was supported by Grant-in-Aid for JSPS Research Fellow, Grant Numbers 15 J10833. I would also like to thank the staff and fellows of the Nanzan Institute for Religion and Culture for the kind assistance during the final stages of completing this essay.

References

Abbreviations

WTZ Watsuji, Tetsurō 和辻哲郎. 1961–63. *Complete Works of Watsuji Tetsurō* 和辻哲郎全集, 20 vols. Tokyo 東京: Iwanami Shoten 岩波書店

WTZ-B Watsuji, Tetsurō 和辻哲郎. 1992. *Supplementary volumes to WTZ* 和辻哲郎全集 別巻, 2 vols. Tokyo 東京: Iwanami Shoten 岩波書店.

Other Works

Crary, Jonathan. 2013. *24/7: Late Capitalism and the Ends of Sleep*. London, New York: Verso.

Dilthey, Wilhelm. 1927. Der Aufbau der geschichtlichen Welt in den Geisteswissenschaften. *Gesammelte Schriften*. Bd. 7. Leipzig, Berlin: B. G. Teubner. English edition: Dilthey, Wilhelm. 2002. "The Formation of the Historical World in the Human Sciences." In *Selected Works*. vol. 3, trans. by Makkreel R. A. and Scanlon J. Princeton: Princeton University Press, 101–209.

Ellul, Jacques. 1954. *La technique ou l'enjeu du siècle*. Paris: Librairie Armand Colin.

Feenberg, Andrew. 1999. *Questioning Technology*. London: Routledge.

Heidegger, Martin. 1954. Die Frage nach der Technik. *Vorträge und Aufsätze*. Pfullingen: Günther Neske, 9–40. English edition: Heidegger, Martin. 1977. "The Question Concerning Technology." In *The Question concerning Technology and Other Essays*, trans. by Lovitt, W. New York: Harper and Row, 3–35.

Heidegger, Martin. 2006. *Sein und Zeit*. Tübingen: Max Niemeyer. English edition: Heidegger, Martin. 1962. *Being and Time*, trans. by Macquarrie, J. and Robinson, E. New York: Harper and Row.

Ihde, Don. 1990. *Technology and the Lifeworld: From Garden to Earth*. Bloomington: Indiana University Press.

Jaspers, Karl. 1933. *Die geistige Situation der Zeit*. Berlin: Walter de Gruyter.

Latour, Bruno. 1994. "On Technical Mediation: Philosophy, Sociology, Genealogy." *Common Knowledge* 3.2: 29–64.

Latour, Bruno. 2002. "Morality and Technology: The End of the Means." *Theory, Culture and Society* 19.5/6: 247–60.

Marcuse, Herbert. 1964. *One-Dimensional Man: Studies in the Ideology of Advanced Industrial Society*. London: Routledge & Kegan Paul.

Verbeek, Peter-Paul. 2011. *Moralizing Technology: Understanding and Designing the Morality of Things*. Chicago: University of Chicago Press.

Watsuji, Tetsurō 和辻哲郎. 1931. *Ethics: The Significance and Method of Ethics as the Study of Human Being* (*Iwanami Lecture Series in Philosophy*) 倫理学:人間の学としての倫理学の意義及び方法 (岩波講座哲学). Tokyo 東京: Iwanami Shoten 岩波書店.

Watsuji, Tetsurō. 1988. *Climate and Culture: A Philosophical Study*, trans. by Geoffrey Bownas. Westport, CT: Greenwood.

Watsuji, Tetsurō. 1996. *Watsuji Tetsuro's Rinrigaku: Ethics in Japan*, trans. by Yamamoto Seisaku and Robert E. Carter. New York: State University of New York Press.

Chapter 4
Watsuji's Ethics of Technology in the Container Age

Thomas Taro LENNERFORS

1 Introduction

A Swedish TV commercial from a major pension company, broadcasting in the summer of 2016, is culminating in the container terminal in a port. An ever growing crowd walk throughout the city, from block to block, until reaching the terminal. They are walking together, towards the better future that would ensue if they would sign up for a pension scheme from this particular company. It is in the middle of the night, and the people are standing beside, between, and on top of the stacked containers. This very presence of flesh, of people, is a complete contrast to how container terminals are usually seen – as being essentially cold and corrugated. A woman is operating the container crane, happy to look out over the unusual sight of a crowd in the terminal. The *persona* of the pension company, a bearded man who goes under the name *Time*, stands on top of a container, waving a torch. Around him, all people are dancing to Michael JACKSON's song: "You are not alone, I am here with you…".

Let us now turn from the container terminal – a stronghold of contemporary technology – to Japanese ethics. When it comes to Japanese ethics, one of the most well-known thinkers who has made his way to the Western context, by accessible translations of major works and by a vivid commentary in the English language, is the Japanese philosopher WATSUJI Tetsurō 和辻哲郎 (1889–1960). Watsuji was mainly active in the first half of the twentieth century, and was an emblematic

The original version of this chapter was revised: MS Mincho font was updated throughout the book for Japanese and Chinese characters. The correction to this book is available at https://doi.org/10.1007/978-3-319-59027-1_13

Thomas Taro LENNERFORS (✉)
Department of Engineering Sciences, Uppsala University, Uppsala, Sweden
e-mail: thomas.lennerfors@angstrom.uu.se

Thomas Taro LENNERFORS, MURATA Kiyoshi (eds.), *Tetsugaku Companion to Japanese Ethics and Technology*, Tetsugaku Companions to Japanese Philosophy 1,
https://doi.org/10.1007/978-3-319-59027-1_4

representative of Japanese ethics, which itself, as is stated in the introductory chapter, can be seen as inherently transcultural (Lam 2011). Watsuji had a long-standing interest in Western philosophy and throughout his life, he teased out connections and distinctions between his own cultural environment and the philosophy growing in it on the one hand, and Western philosophy on the other. Robert CARTER, a main commentator, argues that by reading Watsuji, which in itself is a way to dare to cross traditions, there is much to learn (Carter 2013).

A fundamental standpoint of this paper is that the clearcut distinctions between the East and the West need to be problematized. This is based on a long-standing problematization of which the most famous recent proponent is Edward SAID and his concept of *Orientalism* (Said 1978). Grouping together large populations into the concept of "the Orient" and contrasting it, or even diametrically opposing it to the Occident, is seen as the result of a socio-cultural power dynamic more than anything else, and the differences between the two are often largely exaggerated. Given this tradition of thought, it would follow that the so-called Japanese philosophy is not philosophy only about so-called Japanese people. Rather, it is something that could also be relevant for understanding the global condition. Watsuji's ideas point to something more fundamental than just a representation of the Japanese, and should be read as a philosopher of the human. This follows the claim in the introductory chapter that Japanese philosophy, at least in recent times, is often transcultural. Watsuji is also interested in technology, which is another reason for exploring his ideas. For these reasons, I revisit Watsuji's thoughts in this paper – not to get a Japanese perspective on containerization, but to get the view of an important and interesting philosopher on this issue.

Watsuji has, as mentioned, been recognized in the Western context. One of his most accessible pieces, *Climate and Culture*, was translated to English in 1961, followed by a translation of a part of *Rinrigaku* (Ethics) in 1996. Other translations of Watsuji's work exists, but most of his writings, totalling twenty volumes (Watsuji 1961–1963), are still only accessible to readers of Japanese. This is representative of Japanese philosophy. Apart from the scarce translations, Robert Carter, who played a key role in the translation of Watsuji's *Ethics*, has contributed to much of the formation of how Watsuji is seen in the English speaking world, thanks to his pedagogical and accessible introductions to Watsuji's thought (Carter 1996, 2013). While it might seem as an all-together reckless venture to even try to understand a thinker based on the few books translated into the English language, I also believe that it is important for non-Japanese to engage with the thought of Japanese philosophers, even when the language skills are lacking. Such readers often have a different background than philosophers active in Japan and could possibly identify new aspects, combine Japanese philosophy with non-Japanese practices, or in other ways create new ways of reading and interpreting Japanese philosophy. In this chapter such a project will be attempted.

This purpose of this chapter is to describe WATSUJI Tetsurō's ethics of *ningen* – of human betweenness, explore how it forms the base of an ethics of technology, and analyse a contemporary technology – containerization – based on his ethics. The structure of the paper follows naturally from this purpose. We first need to understand Watsuji's basic view on ethics. Given that this is fairly well-known in the

English speaking context, I have attempted to make the description of this, found in the second part, brief. Furthermore, there is a need to understand Watsuji's ideas about technology – what is its nature and what role did he attribute to it? And how does his view on technology relate to his ethics? This part is not as straightforward as the second part, as most of Watsuji's ethics of technology is implicit in his work. The interpretative work and an attempt to explicitly propose an ethics of technology based on Watsuji can be found in the third part. In the fourth part, to understand containerization, I will embark on a more creative act of transposing Watsuji's thought onto the current technological system. Although at first sight containerization could be seen as concerning only goods, it will be suggested that the principle of containerization goes well beyond the material sphere. Watsuji's ideas are used to analyse both kinds of containerization. This is followed by a concluding section.

2 Watsuji's View on Ethics: Betweenness

From the beginning of our lives we are dependent on other people. We are fed, taken care of, raised, by our parents and other family members. We gradually gain autonomy and a sense of individuality, but always remain part of various communities, including the family. This, I believe, is a description which is applicable in many parts of the world. We are bound up in relations, but are also at the same time individuals. Carter points out that Watsuji sees the human being as fundamentally relational. Carter writes that: "Recognition of our relatedness with others is, in fact, our point of departure on our quest for awareness of our own deeper dimensions, and the goal of our attempts to live a life of personal self-cultivation, in relationship with others, is to 'make our gathering together worth-while'" (Carter 2001: 65). Based on the idea of relationships, Carter explains that "ethics is concerned with those problems that arise between persons" (Carter 2013: 138). This seems almost provocatively simple, but to understand its meaning we need to revisit the original work of Watsuji to delve deeper into the background and more exact meaning of this relational view.

In Watsuji's *Ethics*, which is written in the Meiji Era, a time when Japan had opened up its borders and became more involved in transcultural exchange, Watsuji tried to explain the meaning of the human in Japan in relation to views in the West. He surveyed a number of distinct formulations of the human: the Christian formulation of a being created by God, the reasoned being – *homo sapiens*, the worker or technician – *homo faber*, the person enfeebled by virtue, and the superhuman view "which lifts the self-consciousness of human beings to a higher level" (Watsuji 1996: 13). Watsuji held that all these theories take the *individual* human being as a point of departure. So, although we to some extent know that we are always interconnected in relationships with other people, when we become philosophers, we become increasingly individualist.

In contrast, Watsuji turned back to an exercise in studying the Japanese word for human being, namely *ningen* 人間, which he saw as a very significant word in the Japanese language on the basis of which the Japanese have developed a distinctive conception of the human being (Watsuji 1996: 15). This compound consists of two Chinese characters – person (*hito* 人) and betweenness (*aida* 間). While the first of the two characters is fairly unproblematic, *aida* can mean temporal and spatial intervals, as well as the betweenness between humans. His disciple YUASA Yasuo 湯浅泰雄 explained that: "this betweenness consists of the various human relationships of our life-world. To put it simply, it is the network which provides humanity with a social meaning, for example, one's being an inhabitant of this or that town or a member of a certain business firm. To live as a person means, in this instance, to exist in such betweenness" (Yuasa 1987: 25). (Read further about Yuasa in Chap. 6).

Watsuji's ideas of the relational subject has been connected by Erin MCCARTHY to feminist philosophers in the Western context who see the human being as relational and embodied (McCarthy 2010). Another obvious resemblance is between Watsuji's thought and twentieth century anthropological theory paying great attention to the relational subject (Sahlins 2008).

Carter (2013), in his discussion on Watsuji's influences, explains that in Japanese ethics, especially in Watsuji, the fundamental view of the human being is a double negation. An individual necessarily have to negate society (and his or her human relationships) to become an individual, but at the same time has to negate his or her individuality to become a member of society (and part of human relationships). It is given this very particular way of writing the human in Japanese, something which in itself already contains betweenness, that Watsuji brings forth the human as a combination of the individual and the social:

> What is recognizable here is a dialectical unity of those double characteristics that are inherent in a human being. Insofar as it is a human being, *ningen* as an individual differs completely from society. Because it does not refer to society, it must refer to individuals alone. Hence, an individual is never communal with other individuals. Oneself and others are absolutely separate. Nevertheless, insofar as *ningen* also refers to the public, it is also through and through that community which exists between person and person, thus signifying society as well, and not just isolated human beings. Precisely because of its not being human beings in isolation, it is *ningen*. Hence, oneself and the other are absolutely separate from each other but, nevertheless, become one in communal existence. Individuals are basically different from society and yet dissolve themselves into society. *Ningen* denotes the unity of these contradictories. Unless we keep this dialectical structure in mind, we cannot understand the essence of *ningen*. (Watsuji 1996: 15)

What we find ourselves with is a situation where both the individual and society is negated (but not annihilated), and the commonality that is found is nothingness. As Carter puts it:

> True ethics, then, is a return to authentic unity through an initial contradiction within the self, and between the self and the "other" in the betweenness. The negation of "otherness" reestablishes the betweenness as a nondualistic connection between the self and others that negates any trace of opposition: a negation of a negation. One is now aware of the interconnection of all things, ultimately eliminating any sense of a separate self and other (from within a dualistic perspective). (Carter 2013: 144)

The human being is thus interconnected to other human beings in its very nature. One is born and socialized into ethical relationships and there is no neutral position – a view from nowhere – from which one could think and act. So, rather than having a contemplative view of the human being, Watsuji sees the human being as always engaged in practical action, embedded in the social fabric. Writing a book, what might be seen as pure contemplation, is not a purely intellectual matter, but Watsuji envisions the person sitting behind her desk, typing or scribbling, and later publishing the work. This resounds with turns away from the contemplative subject, towards conceptualizations such as the *vita activa* (Arendt 1958), and the philosophy of practice (Schatzki et al. 2000). This philosophy of life appreciative of pure, immediate experience is most plausibly an influence received from NISHIDA Kitarō 西田幾多郎 (Nishida 1990: xxxiii). Watsuji is further inspired by Confucian ideas about responsibility in various kinds of relationships, but we will not go into that here. Rather, the main gist of this brief overview is the very fact that human nature is betweenness. This betweenness and the relationships between human beings is a fundamental trait of how Watsuji views technology, an issue to which we now turn.

3 Watsuji's View on Technology

The first work by Watsuji that was translated from Japanese to English was *Climate and Culture* (Fūdo) (Watsuji 1988). One argument in the book is the, from today's perspective seemingly anachronistic, argument that climatic conditions shape societies and subjectivities. Carter explains that a person raised in one of the three main climate types – monsoon, meadow, and desert – will have a personality aligned with the climate, and distinct from persons living in the other two major climates. Could this theory be stemming from a Japanese preoccupation with climate, emphasizing the uniqueness of the Japanese climate in comparison to that of other regions – for example the existence of four distinct seasons?

In any case, climate does not only refer to the weather, but inspired by a comment made by Carter in passing, it is a richer concept. Carter writes that "By *climate* Watsuji means to include not only the weather patterns of a region but the natural geographic setting of a people plus the social environment of family, community, society, lifestyle, and even the technological apparatus that supports community survival and interaction" (Carter 1996: 5). Here, one wonders whether the translation of *fūdo* to climate is suitable. A more appropriate translation is *milieu*.[1] Augustin BERQUE argues that like in French, the word *fūdo* translates into the natural and social conditions, and where the scale of this is not specified. A translation into climate naturally leads us to think about environmental conditions and their influences on social systems (Berque 1994: 498).

By prying open the definition and scope of *milieu*, to include the technological apparatus that supports community survival and interaction, we can see that Watsuji is interested in technology, as being one central part of the context or fabric of societies. Although Carter places emphasis on technology, he has not written explicitly about the role that technology plays in Watsuji's work. Nor have I found any other work focusing on Watsuji's view on technology up to this date (see

[1] I am indebted to CHEUNG Ching-yuen and a reviewer to introducing the work of Augustin Berque to me.

however Inutsuka, Chap. 3). In any case, Carter's remark also sets the stage for how Watsuji describes technology – as something which needs to be understood in order to understand a particular society. I therefore argue that Watsuji viewed the study of technology as crucial for understanding the human condition (see also Fors and Laaksoharju, Chap. 9 and Palm, Chap. 10, for discussions about the technology-mediated environment inspired by IMAMICHI Tomonobu 今道友信).

Watsuji held that the particularities of the present and past society were important for the shaping of subjectivities. Watsuji's philosophical work explores the interconnectedness of material practice and mental subjectivities. A philosopher is always immersed in the very fabric of the society in which she lives and through engaging with this society, she could philosophize. Moreover, Watsuji is explicitly having a conversation with German philosopher Martin HEIDEGGER (1962), who Watsuji thinks puts too much focus on time and the individual, while neglecting space and the social dimension of human beings. To pay attention to space is to understand a person's situatedness in a particular *milieu*, and the phenomenological consequences thereof.

With this understanding of technology as part of the *milieu*, we now continue the investigation.

3.1 Technology of and for Ningen

In his *Ethics*, Watsuji discusses newspapers, telephones, radio, traffic, roads, railroads, cars, airplanes, postboxes and the postal system not in order to discuss these particular technologies in themselves, but how they relate to his broader philosophical project. By conceiving the subject as practically interconnected to what surrounds her, it is not surprising that technology becomes a key issue to study. Every day, Watsuji was, and we are, engaging with technological artifacts and systems. And although we often very quickly get used to the technological system surrounding us, Watsuji expresses a fascination with technology, almost re-enchanting it to make it visible and tangible.

Watsuji asks us, in his *Ethics*, to stop for a moment and think about technologies of communication and transportation. He brings up the example of a primitive form of communication and transportation. The simplest form could be that I simply walk to another person and engage in a dialogue. The way to that person is a road, or becomes a road if it is walked several times. This way of relating to the other through open dialogue is an excellent example of the ethical attitude of betweenness that Watsuji described. A second level of technological complexity is when I entrust an envoy to communicate a message. The envoy, the third party, must here understand the content of what is said to be able to communicate this to the other party. A step further is when written words are delivered by a messenger. Here, the messenger does not know the content of the letter but his or her function is just to deliver it. A letter in the postal service removes the need for a dedicated messenger and centralizes the messenger function to a governmental agency. What is shown in this gradual construction of the postal service, is that Watsuji wants to bring down the technological system to our practical everyday life in betweenness.

For Watsuji, technology is therefore a material manifestation of *ningen*, of human betweenness. However, as mentioned in the preceding paragraphs, technology is something that often becomes seen as purely material – as something which exists independently of humans. Indeed, Swedish human ecologist Alf Hornborg (2001), states that the machine, or technology, has risen to become a fetish, something which is ascribed almost magical powers, while it is indeed only one node in a system of human relations. Watsuji makes a similar statement, without the explicitly Marxist approach of Hornborg:

> At first glance, facilities for transportation and communication seem to be objective "things." But they are not mere objects that have nothing to do with subjective beings. Just as gestures or expressions of Thou, who stands over I, are neither motions of physical bodies nor mere vibrations of air but are the relationship of a subjective Thou with an I, so a speeding car, a red mailbox, and a newspaper are not mere physical bodies… In this way, all the "things" constitutive of facilities for transportation and communication are things in the same sense that the human body is a thing, and they are also subjective in the same sense that the human body is subjective. (Watsuji 1996: 164)

It is here argued that in the same way as a human being cannot be divided into mind and body, technology can also not be seen as a mere thing. All technologies are subjective bodies as well, according to Watsuji, since they are always created as a part of society. This resounds of contemporary theories who want to pay more attention to the material world, and how the material world has agency. Of course, scholars such as Bruno Latour (2005) have gone far in developing notions of material agency, but Watsuji can be seen as a part of such intellectual projects about the agentic, performative aspects of technology.

To sum up this first section about Watsuji's view of technology in his *Ethics*, Watsuji claims that technology is artefacts and systems that represent betweenness. They should always be conceived in relation to his concept of *ningen*, human betweenness.

3.2 Technology as an Indicator of Spatiality and Temporality

As has been already mentioned, *Climate and Culture* is aimed at critiquing and supplementing Heidegger's work with a true concern with spatiality. In *Ethics*, Watsuji describes how different forms of transportation and communication technologies unveil not only human betweenness, but also the spatial nature of human beings. Rather than thinking about a hammer, Watsuji considers the technological context in which he is part. Not surprisingly, traffic is a starting point:

> Traffic is a spatial expression of human social intercourse, and traffic patterns gave birth to roads on becoming fixed. Roads stretch out in a spatial manner and are further intersected spatially at certain places. An intersecting point of arteries of traffic expresses intense human social intercourse. (Watsuji 1996: 155)

Traffic is therefore a geographically vast system, which represents the spatiality of human beings. Whether it concerns walking, riding a bike, bus, or train, a fundamental part of the nature of the human is this spatial consciousness. Not only transportation expresses spatiality, but also communication:

> [T]o communicate information to the public signifies that the information is spread spatially, in the literal sense of the term. This is true even in the present world. The most outstanding characteristic of the progress of the communication media, specifically newspapers and radio, lies in strengthening and extending the capacity to spread information spatially. Written words are rapidly carried to every corner of the country by the railroad. Spoken words are spread spatially even more rapidly by means of radio waves. (Watsuji 1996: 155)

These examples of transportation and communication as fundamentally related to space, which are easily understandable, are taken from the everyday practical understanding of persons. Rather than theoretically thinking about some tool and how we relate to it, Watsuji is inspired by the philosophy of life of v and others, starting his philosophy is his lived context, and from thereon proceeds to ontological questions. By his analysis, Watsuji claims that we already understand spatiality in an ontic way (Watsuji 1996: 158), in the way we live in the world, but that we have a harder time understanding the spatiality in theoretical terms: space from an ontological perspective. Often, Watsuji holds, people, in their philosophical investigations, tend to think of the human subject as abstracted from space, that the external, natural, world is spatial, but not the subjective world of human beings. But given his concept of *ningen,* a human subject is already encompassed in a relation between two subjects, that is, it is spatial in itself. He concludes:

> When emphasis is placed on the contention that the essential characteristic of *ningen sonzai* [human existence] is to be regarded not merely as the being of, but rather the betweenness between humans, then we have, in fact, already encountered subjective spatiality. All expressions that indicate the interconnection of the acts of human beings – for example, intercourse, fellowship, transportation, communication – can be understood only with a subjective spatiality of this sort. Spatial extendedness, as is evident in publication, communication, and so forth is an expression of this subjective spatiality. (Watsuji 1996: 157)

To try to explain his point, Watsuji takes the example of the *Great Kantō Earthquake* in 1923, when communication and transportation channels were cut to the affected region. This led to the region being completely separated from the public and the mental relationship between subjects. Therefore, even though the people kept existing in natural space, their connection in mental space was lost (for an analysis of the Great Eastern Japan Earthquake see Chaps. 11 and 12).

Watsuji also points out how technologies of transportation and communication can help us understand the temporality of *ningen.* For example, in the low-tech means of transportation of walking, when we start walking there is beforehand a place and a set of human relationships that we are walking to. Watsuji maintains that walking is the realization of possible human relations (Watsuji 1996: 183). Those human relations, potentiality, represent the future. In engaging in transportation, a temporal aspect is always involved. When will I reach the destination? What will happen? The future is therefore present in every kind of practical activity, the possible betweenness that gives the activity its direction. Watsuji, similarly to his discussion about space, builds up the argument stating that an express train that starts in Tokyo is already bound for Kobe or Shimonoseki, and relationships in the Kansai area, have already served the passengers to board the train (Watsuji 1996: 184). Also, he claims that in a megalopolis, where roads cross, countless numbers of walkers, bicycles, cars, and streetcars come and go, there is also the distinction between the present betweenness and the possible one. Also in these more complex setting, the temporality of *ningen* is expressed through technology (Watsuji 1996: 184–185). As a concluding note, Watsuji holds that

> all of the facilities of spatial connection, that is, the means of transportation and communication, come to have an especially important significance as the expressions of human beings. Without means of transportation or communication, there would be no society. Metaphorically speaking, these constitute the nervous system of society. Their structure reflects the structure of the social existence of human beings, and the historical development of this structure reflects the historical development of human beings. (Watsuji 1996: 160)

The two above sections have concerned how technology is expressing the basic structure of *ningen*. Apart from expressing betweenness, technologies of transportation and communication express the spatial and temporal nature of the human being. Watsuji also claims that technology transforms *ningen*. To this we will now turn.

3.3 Technology as Transforming Society

Going beyond the above reasoning that Watsuji sees technology as a part of the *milieu*, he argues that technology also transforms the betweenness of human beings. For example, by means of advances in transportation technologies, people can relate to each other on a larger geographical basis.

> Thus, the development of a network of railroads gradually transformed the ways of human communication. The intensity of social connections is given expression to by the intensity of railway lines, as well as by the frequency of trains. I am sure that this tendency will be strengthened even further through the emerging facilities of communication such as the automobile and the airplane. (Watsuji 1996: 162)

Here, technology becomes not just a representation of society, but rather as something which might indeed change society, something with an agency in itself.

The well-being of the technological system then significantly maintains or hampers our ways of relating to other beings in society. Watsuji takes the example of postal service. If quick, it can indeed strengthen betweenness. If slow, it can weaken it:

> Under circumstances in which a response to a letter is delivered after a month's interval, we cannot be said to be engaging in a conversation in an active manner. If we receive a response at a time when we have almost detached ourselves from the state of mind we were in while

> writing the letter, then we are unlikely to share the same state of mind. On the other hand, if and when postal services spare no time in delivering words from one person to another both quickly and frequently, then we shall be able to share pleasures as well as pains. A community of beings would thus be realized. (Watsuji 1996: 164)

Although Watsuji is rather silent upon the topic, ideas of how technology shapes society shine through in his examples. Technology has the potential of promoting betweenness through for example the postal service which allows human beings to relate to each other although a large distance separates them. The same goes for other forms of communication technologies and transportation technologies. Technology changes the way we relate to each other, but the forms are dependent on basic patterns of for example transportation and communication, and could therefore be seen as expanded and enhanced forms of such patterns. The forms of interaction are adapted to new forms of technology, but although forms change the basic idea of betweenness is dominant for Watsuji.

3.4 Watsuji's Ethics of Technology

The above subsections have argued that Watsuji sees technology as a contextual, environmental factor, as a part of *milieu*, which represents betweenness, as well as has the potential for strengthening or weakening it. Watsuji does not discuss the ethical implications of technology directly, but it is clear that we must find his view on the ethics of technology within the context of human betweenness.

Given this, one could read Watsuji creatively, and possibly even attribute a normative ethical theory regarding technology. Such a reading, which I attempt here, would hold that Watsuji would see technology which promoted the betweenness of human beings, the technology that furthered the creation of a society, as more appropriate, more ethical. In the examples above, we have seen how Watsuji boiled down the postal system as a complex form of walking to another person and talking to her. This does not, however, mean that he nostalgically sees walking and talking as ethically superior to communicating by letter. Watsuji is separated from the discussion of appropriate technology, later developed by Ernst Friedrich SCHUMACHER, who stated that we need technology in the size of man (Schumacher 1973; Lennerfors 2015). However, Watsuji holds that glitches in complex technological systems, for example when letters are delivered too late, jeopardize betweenness. His attitude towards technological development could therefore be presumed to not be conservative but rather appreciate the complexity of upholding a sophisticated technological society. The simple act of walking and talking are thus much less vulnerable than more complex forms of human communication. The uniting view of these two forms of human communication is betweenness, and this is a main ethical value that technologies should promote.

Furthermore, the view presented here would indicate that Watsuji would not like to see ethics of technology as something fundamentally different than ethics as such. Rather, we should see technology as just one potential promoter of betweenness, and not something that should be discussed in a separate field of scholarly practice. So, the claim of this chapter is that, by drawing out the consequences of Watsuji's work and folding them back upon themselves, Watsuji's ethics of technology could be: to see technology always as a part of betweenness, and to promote technology that build and maintain such betweenness.

Watsuji's contributions towards an ethics of technology are more incidental, the reason of which is that Watsuji probably did not see technology as something which deserved a particular focus, but still was so omnipresent that one could not study ethics without taking technology into account. It is now time to turn to the present, and see what impact Watsuji can have today. What would it mean to be inspired by Watsuji in the present context? How would Watsuji look at the present society?

4 Containerization

In the introduction to this text, we got a glimpse of a Swedish commercial which was staged in a container port and it is now time to get back to discussing containerization. If we turn to Merriam Webster's dictionary for a widely accepted definition we read that a container is (1) an object (such as a box or can) that can hold something, and (2) a large box that goods are placed in so that they can be moved from one place to another on a ship, airplane, train, or truck. While this might have, as container historian Marc Levinson puts it, all the romance of a tin can (Levinson 2006), the introduction and implementation of the container has brought about nothing less than a revolution in transport. For this reason, it should be of interest to Watsuji, who as we have seen above, dedicated significant energy to discuss technologies of communication and transportation, with numerous examples such as roads, cars, trains, airplanes, newspapers, and radio.

An analysis of containerization, given that one follows Watsuji, is to understand how containerization represents betweenness and also affects betweenness. A first objection to the whole project is that containers are used for transporting goods, not humans, and that Watsuji was interested mainly in the transportation of humans. However, Watsuji is also interested in the postal system which does not transport humans, but the messages of humans. In this way, the postal system has the potential to influence betweenness positively in creating and promoting relationships. If we disregard the argument from commodity fetishism, it would not be too bold to suggest that goods are also creations and representations of human beings, and the transportation of goods is just a step in a chain of inter-human relationships. Although relations might seem weak, we are related to people who produce the products we consume, and recently such chains of dependence are stressed more, connecting the consumer and producer. Similar to the fact that the essence of the postal service is not the transport of paper, the global container trade can never be reduced to goods only. To further stress the human aspect of containerization, I will discuss the container metaphorically, to see if we can see any additional implications for betweenness when seen not as a material artifact, but as a broader representation of the present society.

As a final point before moving into the analysis, one could ask why I study the container, rather than other forms of technologies such as social media. The reason is that I was interested in unveiling a very dominant technology that is often hidden from everyday life, but something that I argue is important for understanding the present.

4.1 Containerization of Goods

The standardized container made shipping cheaper. Rather than loading and offloading boxes and bags, cargo could be loaded into ships, trucks and trains in containers. Rather than handling different cargo in different ways, both concerning loading, offloading, and stowage, all cargo could now be handled in the same way. Surely, there are refrigerated containers for fruits, meat, fish, and vegetables, but much dry cargo can be loaded in regular containers. Why is such a simple thing a revolution? Numerous people who had been working in the ports loading and discharging goods were replaced by container cranes, and the time spent in port was also radically reduced, increasing the utilization of the expensive ships. This reduced costs. The container could easily be reloaded on trains or trailers, and made shipping increasingly intermodal. The corrugated steel always protected the goods inside against bad weather, bad handling, and theft, which were always significant problems when it came to cargo handling. What is obvious is that this technological shift has revolutionized the way we transport goods across sea and land. This has also led to an increased number of goods to low prices being available to consumers, a key component in what we see as globalization.

While the global container trades could on the one hand be seen as bringing people closer together, allowing exchange of goods and ways of life, thus promoting betweenness, there are of course many critics of global trade. Such trade is often embedded in terms of trade that favor some actors over others. Containerization, as a way of promoting global trade, is then a facilitator of such relationships, and by accelerating global trade, it can also increase inequalities in the economic system.

If Watsuji would have looked at containerization, he would have unpacked it and made it visible to his readership. If containerization and the underlying global trade flows are not discussed, it is easy to fall into commodity fetishism, believing that goods exist irrespective of human beings and human labour, readily available in supermarket stores. By focusing on the transport of goods, we can begin to ask some very basic but important questions: Where was this good produced and by whom? How did it come here and who was involved in its transport? Container technology can therefore be a way to form an understanding of the human relationships that are the backbone of global trade. In this sense, containerization represents betweenness. Containerization also affects betweenness in accelerating and intensifying the flow of goods between human beings. If such trade is by tradition unequal, meant to serve the interests of those in the core of the world system (Hornborg 2001), containerization promotes such development. But rather than seeing technology as the culprit, it is more of an accelerator to already problematic relationships.

4.2 *Containerization of People and Thought*

While the previous part dealt with the transportation of goods, both potentially promoting betweenness since the fruits of labour are shared across the world, and unequal exchange in the world system, I will now turn to a less material take on containerization, inspired by the 2015 book *The Container Principle* by Alexander Klose. In the book, the author puts new perspectives on the container. The book goes beyond economic theory, and sees the container from the light of philosophy and social theory. It thereby unleashes more radical thoughts that are of relevance to this paper, in that they ask us what the container represents, and ask us how we are affected by the container *principle*, rather than the container. Two examples from Klose's book will show how different his approach is from for example that of Levinson, seen in the preceding part. First, Klose uses the example of containers as time capsules. Container ships sink and leaves numerous containers spread over the ocean floors. They could potentially keep the goods intact and function as time capsules, the content of which might be unraveled in the future. Second, Klose shows how containerization goes beyond transport, which is represented by modular containerized data centers. The number of data center containers can be decided by the needs of the particular location. One could also use furnished containers as flexible houses, which can be moved easily to any location where they are needed.

But Klose does not stop with the preceding examples. He goes on to argue that containerization is a representation of society in a more fundamental way. The first way to go beyond the material, and see the container as a metaphorical representation of our society, is contrary to common sense to get as close as possible to the material realities of the container. A container is individually integral and enclosed, consisting of a safe hard shell protecting the content from any contamination or influence. One could even venture to argue that in a society strongly influenced by containerization, even individual human beings are containerized. This goes hand in hand with a theory about the human subject where humans by projecting "their container-like self-perception onto the objects around them, the subjects separate themselves as closed-off, container-like entities from the container-like objects in their environment" (Klose 2015: 67). In the theory of containerization of human beings, we find individuals as integral, closed-off, self-sufficient, untouchable, and isolated. Furthermore, we might argue, containers are standardized, and can be dealt with in a standardized way. This description resounds of Boltanski and Chiapello's (2005) theorizations of the perfect employee in the new spirit of capitalism, an employee who could fit into any context, something we could even call an "intermodal employee." The containerization of the employee would mean a standardized person, fitting into a stack of other container-persons, on top, under, or next to. The employee would be replaceable if broken, easily substitutable by any other.

Containerization of people and employees, but isn't society in itself also containerized? While on the one hand, new communication technologies connect us all in the proverbial global village, and makes us exposed to news happening in any part of the world, such technologies are also containerizing us. Social media is often argued to induce more silo thinking, and lead people to be more fixed, hard-shelled, integral, than what would be desirable in a society that would promote betweenness. Targeted news and targeted ads that find us, also strengthens the perception of who we are, making us even more individualized, and individuated. Presentations of ourselves at the internet might also be more containerized in the sense that all we can see is the corrugated steel – the surface. Maybe, as the principle of containerization would tell us, globalization fuelled by ICT is, rather than creating a global village, creating a multiplicity of container societies; small, integral, islands of homogeneity, small villages, rather than a global village. Maybe, containerization affects our thought in that we sort things and people into boxes, rather than seeing the connections and betweenness of people.

So, in short, what could the containerization principle tell us? First of all, in a goods specific way, it tells us that human interactions are intensified by the efficiency of goods transports. This could be promoting betweenness, in the same way as Watsuji's example of the numerous connections between human being in a metropolitan area. It could create betweenness since the goods are shared across the world, but could also contribute to promoting unequal exchange and a less equal relationship between human beings. From a more metaphorical perspective, containerization leads to individuals being enclosed, protected, sealed-off. This would go very much against Watsuji's view of what constitutes the human being – namely to be a *ningen*. My reading of the commercial that was presented in the introduction is that containerization is a fundamentally lonely process. The reason why people are singing "you are not alone" and why the container port is used as a locale for the commercial, is that containerization represents the most individuating process that might exist. The singing of "you are not alone" reminds people that they are not alone, although everything points in the other direction.

5 Conclusion

Not much has been written about Watsuji in the context of technology. We can conclude that Watsuji was interested in technology in his writings, since he saw this as a fundamental part of the *milieu*. However, he did not develop an ethics of technology as such. Watsuji rather uses technology in his writings to think about the human condition – technology says something about human beings and the society they live in. Technology is representing betweenness and can also promote betweenness. For example, the technological apparatus of the postal system is both representing and promoting betweenness. It represents it since the seeming transport of paper is indeed messages that bind people together. It promotes betweenness by intensifying human relations over large spaces, when the technology works well.

While technology represents betweenness, I have argued that one could read Watsuji more normatively as arguing that technology should promote betweenness. In consequence, technology which fails to do so, or technology which once has promoted in-betweenness but later degenerates, should be reformed. This is Watsuji's inexplicit ethics of technology in a nutshell.

We can use Watsuji's thought in two ways. First, to follow him in the analysis of contemporary technology to see how it represent our society, particularly with respect to betweenness. Second, to see if and how the technology of the contemporary situation promotes betweenness or not. This would be Watsuji's contributions to analyses of contemporary technology. Writing more than 50 years after Watsuji's death, this text places its emphasis on containerization, which has revolutionized shipping since the 1960s. I enlarge the framing of the container to be a metaphor of the society in which we are living. By doing this move, containerization of people and not of goods might tell us that we are living in an era of increasingly individual, standardized, hard-shelled boxes, which goes against the idea of relationality and betweenness that Watsuji is promoting. A more normative agenda based on Watsuji's work would be to counter containerization, maybe not of goods, but of people and thought. Rather, what could be promoted is a renewed focus on relationships between human beings. *Ningen* could therefore be one important norm for all analyses on past, contemporary and future technology.

Before ending this piece, with this slightly sceptical reading of containerization, the reader might have made a more positive tone which subsists through this piece. When considering it, the fundamental characteristic of a container is not only the outer shell but also its emptiness. Thus, the container is an empty space, a potential space. Although I have in this paper focused on the hard shell, could this emptiness not be a productive point of departure for thinking containerization in a more positive way? Extending the argument of double negation, a container could be the representation of the void par excellence, the void which is the basis of pure experience, the void which one returns to after having negated individuality and sociality, the void which is central to Japanese philosophy. Such view of the current state of containerized society could even give a hope for a multiplicity of ways in which society could develop. As Carter phrases it:

> Nothingness also signifies the importance of our becoming *empty*, for Watsuji, where we leave the ego-self behind and enter into a relationship with much, or all, of our psychological baggage excluded. From this state we may now enter the nothingness between us as an empty vessel, and if the other is also empty (of their prejudice, habits, and fears), then a genuine relationship is highly likely. (Carter 2013: 169)

Containerization could thus represent a hope that there could be a self, a container-self even, that is intact and integral irrespectively of its content. An empty vessel, an open self, which has left its baggage and ego behind.

References

Arendt, Hannah. 1958. *The Human Condition*. Chicago, IL: University of Chicago Press.

Berque, Augustin. 1994. "Milieu et logique du lieu chez Watsuji." *Revue Philosophique de Louvain* 92.4: 495–507.

Boltanski, Luc, and Eve Chiapello. 2005. *The New Spirit of Capitalism*, trans. by Gregory Elliott. London/New York: Verso.

Carter, Robert E. 2001. *Encounter with Enlightenment: A Study of Japanese Ethics*. New York: State University of New York Press.

Carter, Robert E. 2013. *The Kyoto School: An Introduction*. New York: State University of New York Press.

Heidegger, Martin. 1962. *Being and Time*, trans. by John MacQuarrie, and Edward Robinson. London: SCM Press.

Hornborg, Alf. 2001. *The Power of the Machine: Global Inequalities of Economy, Technology, and Environment*. Walnut Creek, CA: AltaMira Press.

Klose, Alexander. 2015. The Container Principle: How a Box Changes the Way We Think. Cambridge, MA: MIT Press.

Lam, Wing-keung. 2011. "The Making of "Japanese Philosophy": Nishi Amane, Nakae Chōmin and Nishida Kitarō." In *Whither Japanese Philosophy? III Reflections through Other Eyes* (UTCP Booklet 19), edited by Takahiro Nakajima. Tokyo: UCTP, 69–80.

Latour, Bruno. 2005. Reassembling the Social: An Introduction to Actor-Network Theory, Oxford: Oxford University Press.

Lennerfors, Thomas T. 2015. "A Buddhist Future for Capitalism? Revising Buddhist Economics for the Era of Light Capitalism." *Futures* 68 (Apr. 2015): 67–75.

Levinson, Marc. 2006. *The Box: How the Shipping Container Made the World Smaller and the World Economy Bigger*. Princeton, NJ: Princeton University Press.

McCarthy, Erin. 2010. *Ethics Embodied: Rethinking Selfhood Through Continental, Japanese, and Feminist Philosophies*. Lanham, MD: Lexington Books.

Nishida, Kitarō. 1990. An Inquiry into the Good, trans. by Masao Abe, and Christopher Ives. New Haven, CT: Yale University Press.

Sahlins, Marshall. 2008. *The Western Illusions of Human Nature*. Chicago, IL: Prickly Paradigm Press.

Said, Edward W. 1978. *Orientalism*. New York: Pantheon Books.

Schatzki, Theodore R., Karin K. Cetina, and Eike von Savigny, eds. 2000. *The Practice Turn in Contemporary Theory*. London: Routledge.

Schumacher, Ernst F. 1973. *Small is Beautiful – Economics as if People Mattered*. New York: Harper Perennial.

Yuasa, Yasuo. 1987. *The Body: Toward an Eastern Mind-Body Theory*, edited by Thomas P. Kasulis, trans. by NAGATOMO Shigenori and Thomas P. Kasulis. New York: State University of New York Press.

Watsuji, Tetsurō 和辻哲郎, 1961–63. Complete Works of Watsuji Tetsurō 和辻哲郎全集, 20 vols. Tokyo 東京: Iwanami Shoten 岩波書店.

Watsuji, Tetsurō. 1988. *Climate and Culture: A Philosophical Study*, trans. by Geoffrey Bownas. Westport, CT: Greenwood.

Watsuji, Tetsurō. 1996. *Watsuji Tetsuro's Rinrigaku – Ethics in Japan*, trans. by YAMAMOTO Seisaku and Robert E. Carter. New York: State University of New York Press.

Chapter 5
MINAKATA Kumagusu – Ethical Implications of the Great Naturalist's Thought for Addressing Problems Embedded in Modern Science

SATO Maki

1 A Brief Introduction to MINAKATA Kumagusu

MINAKATA Kumagusu 南方熊楠 was born in Wakayama 和歌山 prefecture in 1867 and died in 1941 (Tsurumi 1981; Karasawa 2015). Although he published papers and wrote tremendous amounts of notes and letters, only three books were published during his lifetime. In spite of that, he is well known as the first Japanese scientist to have his paper published by *Nature*: dated 13 October 1893, titled "The Constellations of the Far East" (Minakata 1893).[1] During his youth, he travelled to many countries including Cuba. He left Japan in 1886 at the age of twenty to study abroad, first in the US and then in the UK. Kumagusu is noted for his broad range of interests, not only in his biological studies on fungus and myxomycete (slime mould/*nenkin* 粘菌) which he mainly focused on, but is also well known for his research in natural history studies, Japanese folklore anthropology studies, and in his social activity to protest against the Shrine

The original version of this chapter was revised: MS Mincho font was updated throughout the book for Japanese and Chinese characters. The correction to this book is available at https://doi.org/10.1007/978-3-319-59027-1_13

[1] For further information on Kumagusu's contribution to *Nature*, refer to Īkura (2005).

SATO Maki (✉)
University of Tokyo, University of Tokyo Center for Philosophy,
Graduate School of Arts and Sciences, Meguro, Tokyo, Japan
e-mail: maki.sato@utcp.c.u-tokyo.ac.jp

Thomas Taro LENNERFORS, MURATA Kiyoshi (eds.), *Tetsugaku Companion to Japanese Ethics and Technology*, Tetsugaku Companions to Japanese Philosophy 1,
https://doi.org/10.1007/978-3-319-59027-1_5

Merger Order (*jinjagōshirei* 神社合祀令) in trying to protect the shrine forests (*jinjarin* 神社林) in his later life.[2]

Kumagusu was unfortunate in the academic field since he could not get a satisfying position in a Japanese university. However, his research on myxomycete leads him to have an opportunity to present a lecture for Emperor Shōwa 昭和天皇 in 1929. He left a huge amount of notes, letters, and scribblings which are still under investigation by Japanese scholars and researchers in his hometown Tanabe 田辺 in Wakayama prefecture.[3] Throughout his life, Kumagusu loved and devoted himself to science and the scientific approach (empirical-fact-based analysis on the relation of cause and effect) in analysing his research subjects. However, because of his devotion and enthusiasm for science, he was also quick to notice and find its limitations.

Kumagusu's main argument in pointing out the problems embedded in science and its approach could mainly be found in his correspondences with a *Shingon* school Buddhist priest DOGI Hōryu 土宜法龍 (Minakata 2010: 14),[4] which began in 1893 and lasted till 1922, a year before Hōryu passed away (Minakata 2010: 359).[5] Their key discussions regarding science are contained in their correspondences from quite an early stage during the period from 1893 to 1916. In those letters, Kumagusu made an attempt to challenge the scientific approach by utilizing the idea of mandala[6] from the *Shingon* school. In doing so, he also indicated limitations regarding causality based analysis and the linear concept of time embedded in the scientific approach.

In his letters, he drew a series of his own mandala, popularly known as *Minakata Mandala* named by TSURUMI Kazuko 鶴見和子 (1918–2006). However, since he occasionally called himself *Kinzoku Ō Nyorai* 金粟王如来 in his letter to Hōryu – Kumagusu calling himself as *Kinzoku Ō Nyorai* first appeared in his letter dated

---

[2] The Shrine Merger Order issued in 1906 by the Meiji government was an Imperial Order (*chokurei* 勅令) to merge or to abolish the shrines with a purpose of strengthening the National Shintō (*kokka shintō* 国家神道). The movement continued till 1914 till 35 percent of the shrines (approximately 70,000 shrines) were abolished or merged into a larger shrine. Kumagusu took a lead in the movement against the Shrine Merger Order in order to preserve traditional customs and practices, and to protect the nature, mainly shrine forests, surrounding shrines. His archived statement letter, opposing to the Order, is mainly addressed to Prof. SHIRAI Mitsutarō 白井光太郎 of Tokyo Imperial University.

[3] Kumagusu's house is preserved with his writings at Minakata Kumagusu Archive (Minakata Kumagusu *Kenshōkan*).

[4] DOGI Hōryu (or TOKI Hōryu/TOKI Hōrin, 1855–1923) was a *Shingon* 真言 school priest. He studied at the Keiō University with SHAKU Sōen 釈宗演 (1859–1919) who was a *Rinzai* 臨済 priest under the guidance of FUKUZAWA Yukichi 福澤諭吉 (1835–1901). In 1893, both Hōryu and Sōen were invited to give a speech during the World's Parliament of Religions held in Chicago in September. On his way to Paris after the Chicago conference, he stopped by at London, where he met Kumagusu at a home party held on 30 October 1893 at NAKAI Hōnan 中井芳楠 (1853–1903)'s house. Kumagusu spent with Hōryu from 31 October till 3 November, till when Hōryu had to depart from London to Paris.

[5] The correspondences between Minakata Kumagusu and Dogi Hōryu begun on 19 December 1893 and was understood that it ended on 8 May 1916 with Kumagusu asking Hōryu to send him back all his letters sent to Hōryu. However, a recent research conducted, in 2004, has discovered that their correspondence started on 3 November 1893 and continued until 1922, a year before Hōryu passed away.

[6] According to the Oxford English Dictionary, "mandala" originally means "disc" in Sanskrit, and is a circular figure representing the universe in Hindu and (mantra) Buddhist symbolism.

21 December 1893 (Minakata 1971: 154, 245–276) – I personally think he would have preferred his mandala to be named *Kinzoku Ō Nyorai Mandala* instead of *Minakata Mandala*.[7] In *Minakata Mandala* he pointed out that "causality" connotes the conceptual difference of "causality (*inga* 因果)" and "dependent co-arising (*engi* 縁起)". Furthermore, he criticised the limitation in the western scientific view grounded on the linear conceptualisation of time as a key notion.[8] He mainly argued how modern science is focusing only on investigating "causality," which is based on the idea of a linear concept of time implying the continuity of a cognized event through past, present, and future. Accordingly, he points out that with this linear concept of time, modern science may be ignorant of the complexity of the environmental system as a whole. Such a scientific view encourages humans to observe the environmental system only through occurred events as objects to be scientifically analysed based on chronological change which is detectable by humans. However, he holds, it is not possible for humans to understand the complexity of the whole universe through those scientifically identified and analysed events. In this chapter, through Kumagusu's letters to Hōryu, I would like to draw out some ethical implications that could still be applied to our contemporary science and technology through reading Kumagusu's writings.

2 Kumagusu's Challenge to Modern Science

2.1 Concepts of "Event 事," "Heart-Mind 心," "Substance 物," and Time

Kumagusu expresses his criticism of science from quite an early stage in his correspondence with Hōryu, referring to the three aspects of the universe. He identified the three aspects as "heart-mind (*kokoro* 心)", "substance (*mono* 物)", and "event (*koto* 事), "[9] According to his explanation, an "event" occurs when the "heart-mind" and "substance" interact with each other, and the "event", too, has its own "causality."

> When I refer to the "studies on the event (*kotonogaku* 事の学)" [...] I am referring to the action (or effect) that occurs when the "heart-mind world (*shinkai* 心界)" and "substance world (*bukkai* 物界)" interferes (mixes, or interacts...)... The "event" is known as it has its own causality (*inga* 因果).[10] (21 December 1893) (Minakata 1971: 145)

---

[7] *Kinzoku Ō Nyorai* (the Golden Grain Tathagata) originally appears in the *Vimalakirti Sutra* (*Yuima kyō* 維摩経). *Kinzoku Ō Nyorai* is known as the other name of the layman, *Yuima* 維摩 in *the Vimalakirti Sutra.*

[8] In Kumagusu's writings, he often indicates the differences between the West and East. However, it is not my intention to follow his simplified view on the differences of the West and East, neither to show such differences in this chapter.

[9] *Mono* 物 is translated as a "substance", not as a "matter", since I wanted to emphasize Kumagusu's intention on criticizing that the science focuses on the limited aspects of the research object. *Kokoro* 心 is translated as "heart-mind" and it connotes "consciousness" as well. I chose "heart-mind" with an intention to reflect his interests on dreams and spiritual ghosts in his later life.

[10] All the Japanese text translations from here onwards are done by the author. See Quotation 1 for Japanese.

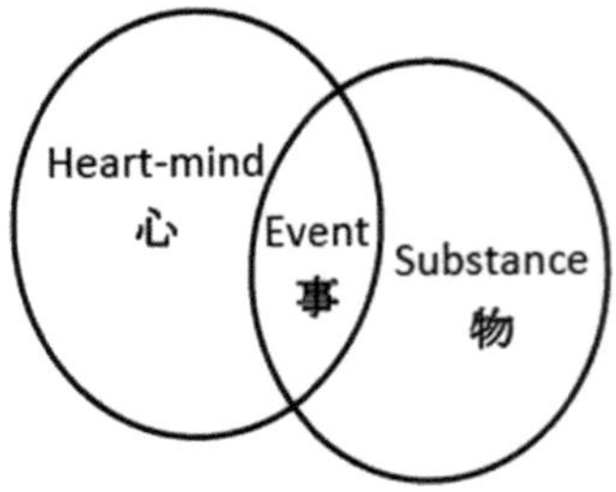

Fig. 5.1 The Three Aspects of the Universe. (Figure 5.1 is a drawing by the author. The original drawing of Kumagusu can be found in Minakata (1946: 15))

Kumagusu thinks that the limitations of science lie in Western scholars since they mainly focus on studying "heart-mind" and "substance". He thinks that by looking into the causality of the "event" itself and by initiating "studies of events" he can identify the "relational actions" happening between "heart-mind" and "substance" (*shinbutu ryōkai renkan sayō* 心物両界連関作用), and its causal effect, that is the continuous chains of causing "events" (Minakata 1971: 148). What Kumagusu means here by "heart-mind" is a function of "substance"; "event" as a phenomenon that is detected through the humans' cognitive capability that can be recognized as a part of a research object. Since Descartes, science has followed the subject-object divide to analyse and to objectify the objects in question. However, Kumagusu thinks that the limitation of science lies in the very divide of subjects and objects. Therefore, he argues for the importance of scientific analyzation on the triangular relationship of "heart-mind," "substance" and "event."

His original drawing (Fig. 5.1) does not describe the endless chain of causal effects between "heart-mind," "substance" and "event" as he describes in his explanation. When his description is precisely drawn in a figure, it can be drawn as Fig. 5.2 (Minakata 1971: 145–148). Kumagusu thinks that first, with a mutual interaction of "heart-mind" and "substance," an "event" occurs – this could be understood as a reason he drew his figure (see Fig. 5.1) as in a Venn diagram, making the "heart-mind" and "substance" overlapping with each other where an "event" occurs. After the occurrence of an "event," there is a feedback loop from "event" to "heart-mind," which initiates the mutual interaction of "heart-mind" and "substance." According to Kumagusu, this continuous chain reaction (or domino effect) causes continuous "reactions" which can be detected by humans as phenomena. He names this function as *shinbutsu ryōkai renkan sayō*; the "relational actions" happening between "heart-mind" and "substance".

> Scholars nowadays (most of the scientists and the European philosophers) discuss only either on "heart-mind (*kokoro* 心)" or "substance (*mono* 物)." I would like to study with a focus on "event (*koto* 事)," which occurs when the "heart-mind" and "substance" interact (which could also be seen as a phenomenon in the human world), and would like to know the differences and similarities between the "heart-mind world (*shinkai* 心界)" and the "substance world (*bukkai* 物界)."[11] (21 December 1893) (Minakata 1971: 146)

In this passage, Kumagusu clearly recognizes and criticises the limitations embedded in the scientific approach. Through his argument on the "event" as a phe-

[11] See Quotation 2 for Japanese.

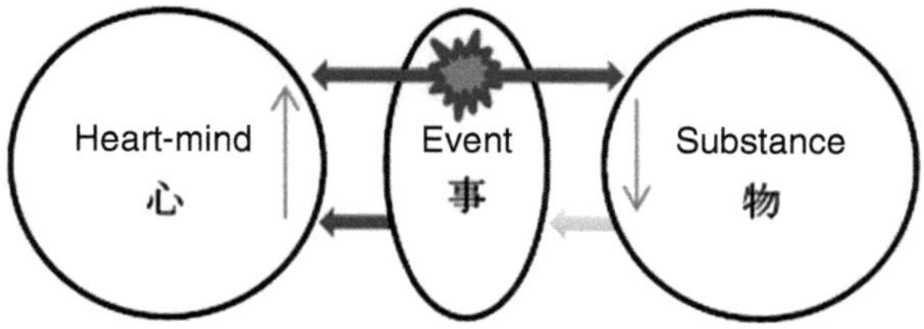

Fig. 5.2 The Chain Reactions of Heart-mind, Substance and Event. (Figure 5.2 is an original drawing by the author)

nomenon, he is trying to point out that there are research objects, which cannot be thoroughly analysed by science and its method of objectifying the "heart-mind" and "substance." In other words, he points out that science is yet to analyse the "world (or the sphere/*kai* 界)" of the research objects and its relationship. By pointing out the "world" of the research objects, he indicates here of a huge void which is ignored and externalized from modern science which focuses mainly on the research object itself. He argues that the scientific approach ignores how the "world" of the "heart-mind" and "substance" interacts with each other and suggests a sphere where actual "heart-mind" and "substance" reside. Specifically, Kumagusu supposes there exists a "world" seen from the object itself that differs from the world understood by the humans.[12] Although this argument sounds similar to the Cartesian duality of matter and mind that influences each other to some extent, Kumagusu's argument is different in the sense that it mentions a "world" beyond the human cognition of the object.

Moreover, regarding the chronological concept, Kumagusu indirectly criticises the simple linear concept of time, commonly shared in the Western scientific field. That past is followed by present and present unfolds itself into future which indicates the irreversible arrow of time. Rather he thinks that such linear concept of time is limited to human perception and argues for a different concept of time in need.[13] Kumagusu seemingly developed his unique concept of time from his reading of *Avatamska Sutra* (the Flower Garland Sutra/*Kegon gyō* 華厳経) and its related thought combined with *Shingon* esoteric Buddhism (Nakazawa 2006: 373–388).[14]

[12] Kumagusu's idea of "world/ *kai*" resembles the "theory of *Umwelt*" developed by a biologist, Jakob von Uexküll (1864–1944). The "theory of *Umwelt*" is developed based on the studies on how various organisms sense their world. In other words, Uexküll's attempt was to overcome the subject-object divide by focusing on the "environment (*umwelt*)" constructed by the organisms (research object).

[13] The differences between objective time (clock time) and subjective time is well known among the scholars in psychology and behavioural science. The latest quantum mechanics works on explaining the different time notion and on different concepts of time that might scientifically be explained logically.

[14] It is interesting to note, as Nakazawa Shinichi analysed, Kumagusu has developed his own reading and understanding of the sutras by coupling the teaching of the *Avatamsaka Sutra* and the *Mahavairocana Tantra* (*Dainichi kyō* 大日経). The author is aware that the Vairocana in *Avatamsaka Sutra* and Mahavairocana in *Mahavairocana Tantra* have different origins. However, in the manuscripts of Kumagusu, his understanding regarding the difference between Mahavairocana and Vairocana is not clear. In his letter written on 3 March 1894, Kumagusu describes Mahavairocana (*Dainichi nyorai*) as a "strange and superb wonder (妙不可思議がすなわち大日如来なり)" indicating the primordial Buddha. Therefore, here, I chose to translate *Dainichi nyorai* 大日如来 as Mahavairocana instead of Vairocana to stress the unique reading of Kumagusu.

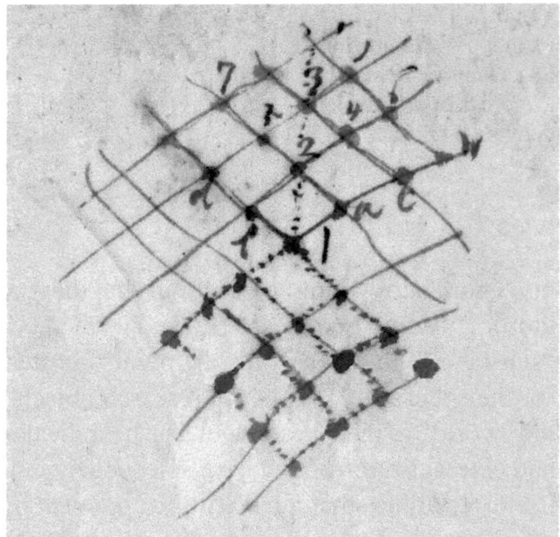

Fig. 5.3 Mesh-like Interconnectedness of the Universe. (Copyright authorized by Minakata Kumagusu Archive (Minakata Kumagusu Kenshōkan))

From his reading of the *Avatamsaka Sutra*, Kumagusu claims that for Mahavairocana (*Dainichi nyorai* 大日如来) there exists only "present," and there is neither "past" nor "future." Kumagusu writes to Hōryu that he learned about Mahavairocana and *Shingon* (in his words, *Dainichi shingon* 大日真言) through his parents' teaching (Minakata 1971: 155–156). Therefore, it could be assumed that he developed his own unique interpretation on the sutras afar from the orthodox Buddhist school teaching. In his letter dated 1 March and 3 March 1894, he wrote his own view of Buddhism arguing that the origin of Buddhism does not come from the Buddha, but Mahavairocana itself. The traces of Kumagusu's reading *Avatamsaka Sutra* could be found from his second letter to Hōryu onwards.[15] By drawing a mesh-like figure (Fig. 5.3) and a neuron-like figure (Fig. 5.4), which depicts "Indra's jewel net," he explains his concept of time by explaining how the teachings of Shakyamuni spread out through time.[16] His concept of time spreads out, weaves, unfolds and interacts like lacework between the present, past, and future.

> As shown in the above drawing of the mesh (see Fig. 5.3), the two gather into one, one spreads into two, and it doubles. However, actually it is not even a doubling, but as shown in the drawing below (see Fig. 5.4), like knitted lacework, a hundred gathers into one, and one spreads into a hundred.[17] (3 March 1894) (Minakata 1971: 210–212)

By utilizing the Indra's jewel net concept of the interconnectedness of the universe found in the *Avatamsaka Sutra*, Kumagusu applies the similar drawings of Indra's jewel net to explain his unique conception of chronology. Though not written explicitly by Kumagusu himself, it can safely be assumed that his concept of

[15] For further reference see Minakata (1971: 175 (Letter dated 21 December 1893), 200 (5 February 1894), 228 (4 March 1894), 392 (8 August 1903)).

[16] "Indra's jewel net" is developed from *Avatamsaka Sutra* and is normally used to describe the interconnectedness of the universe, inspired by a jewel net and a reflection in a jewel, reflecting endless reflections of other jewels (Cook 1977).

[17] See Quotation 3 for Japanese.

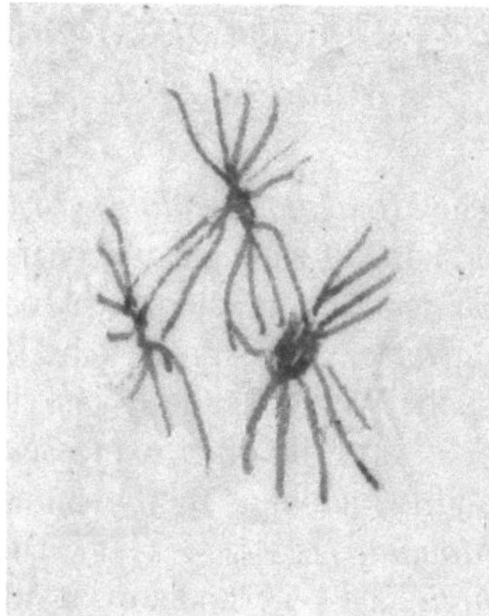

Fig. 5.4 Neuron-like Interconnectedness of the Universe. (Copyright authorized by Minakata Kumagusu Archive (Minakata Kumagusu Kenshōkan))

time is different from the one shared by Western scientists. Science was able to develop through analysis of a partial phenomenon, focusing only on those aspects which are recognisable to humans as a research object, based on the imagined concept of linear time with a pre-set hypothetical "start" ("initial") point and an "end" point of the phenomenon. In Kumagusu's words, scientists are only acknowledging "event" as a fruit (or a result/*ga* 果) of relational actions (or cause/*in* 因) between "heart-mind" and "substance," Kumagusu's conceptual framework does not draw a simple straight line running from cause (past) to effect (present) – with the starting point and the ending point clearly specified – as that would be the laboratory experiments in general, where the analysis of a chronological change of a substance in a given condition is studied as a basic approach to scientifically understand a substance under analysis. Rather, Kumagusu's chronological concept has no start and no end, it continues on and on through long spans of time, weaving a complex net of time. This unique conception of Kumagusu chronology derived from his attempt to explain his understanding of the origin of Buddhism that Buddhist philosophy must have existed before the Shakyamuni. In other words, Kumagusu thinks that Shakyamuni alone could not have come up with all the teachings, but the very existence of Mahavairocana must have in some way inspired Shakyamuni, that Shakyamuni's teachings are the "result" of the "cause," Mahavairocana.[18]

In consequence with these two drawings, as a counter-idea to the Western science based on the linear concept of time, Kumagusu describes that causality only identifies the reason (*jōri* 条理) of the "event" and gives an example of how physics identifies the causal relation of the "substance" by reflecting on electricity generation. However, he thinks that humans cannot identify the fundamental reason that penetrates the universe through understanding causal relations. Additionally, he explains that without further understandings of the relationship between how the "heart-mind" and the "substance" interwork together, the "event" observed could only be understood through causality.

[18] I am not going to discuss into details of this Kumagusu's understanding of Buddhist teachings in this essay since I believe it is another issue to be argued in a different occasion.

2.2 Concept of "Wonder (Fushigi 不思議)": Minakata Mandala op.1

The *Minakata Mandala op.1* (see Fig. 5.5) which is a well-known and well-studied drawing of Kumagusu appears in his letter to Hōryu, written on 18 July 1903. In the *Shingon* school, there are two main mandalas which symbolise the two realms of the universe. One expresses the *Vairocana Tantra* (*Dainichi kyō* 大日経) and is known as the Womb Realm (Sanskrit: Garbhadhatu; Japanese: *Taizōkai Mandala* 胎蔵界曼荼羅). The other expresses the *Vajrasekhara Tantra* (*Kongōchō kyō* 金剛頂経) and is known as the Diamond Realm (Sanskrit: Vajrhadhatu; Japanese: *Kongōkai Mandala* 金剛界曼荼羅). These two mandalas are always displayed in a set (*nini-funi* 二而不二) and are considered to be an expression of the entirety of the Dharma (Minakata 1971: 385). However, *Minakata Mandala op.1* has no clear relationship with the *Shingon* school nor their mandalas, and TSURUMI Kazuko coincidentally named his drawings as Minakata Mandala for convenience (Tsurumi 1981: 82).[19]

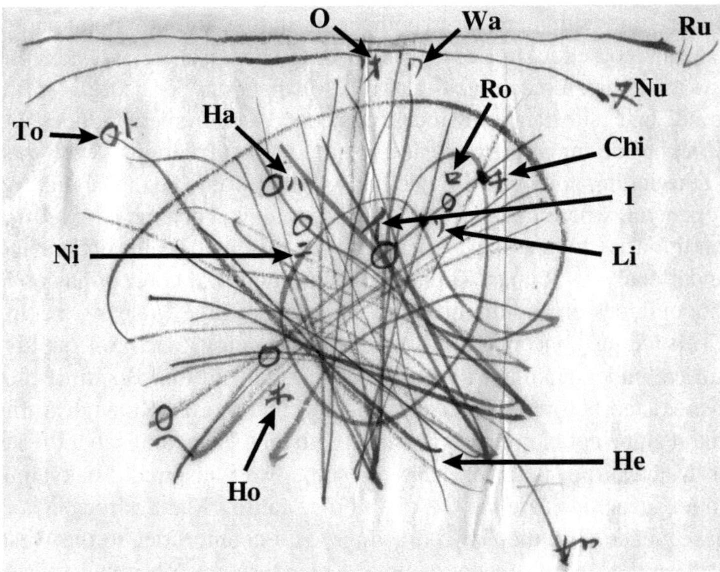

Fig. 5.5 Minakata Mandala op.1. (Copyright authorized by Minakata Kumagusu Archive (Minakata Kumagusu Kenshōkan))

[19] Kumagusu himself acknowledges that he is drawing mandala in his letter to Hōryu dated 8 August 1903 and 24 March 1904, pointing to his drawing in his letter dated 8 August. To clearly differentiate the two, I have named the drawing dated 8 August as *Minakata Mandala op.2*. As mentioned, the vernacular term, *Minakata Mandala*, was coined by TSURUMI Kazuko in reference to Kumagusu's drawing in his letter written on 18 July 1903 (I have named it, *Minakata Mandala op.1*). However, there is no sign or writings left that Kumagusu himself thought *Mianakata Mandala op.1* as his mandala. Tsurumi confesses in her book that she got the idea of calling Minakata's drawing, *Minakata Mandala* from Dr. NAKAMURA Hajime 中村元.

Nine years after Kumagusu drew his concept of the relation between "heart-mind", "substance", and "event" (Fig. 5.1) based on the idea of infinite time (time with no beginning and no end; Figs. 5.3 and 5.4), he further elaborated his idea and drew *Minakata Mandala op.1* to explain the limit of the Western scientific approach, touching upon his concept of "wonder (*fushigi* 不思議)."[20] I consider Kumagusu as having tried to explain the hidden or undetected "unknowable knowledge (*fukachi* 不可知)" behind the identified scientific facts, through the drawing of *Minakata Mandala op.1* and its related explanation.

The following is his explanation of "wonder", which divides wonder into five wonders all belonging to Mahavairocana:

> There are wonders (*fushigi* 不思議): "event wonder (*koto fushigi* 事不思議)," "substance wonder (*mono fushigi* 物不思議)," "heart-mind wonder (*kokoro fushigi* 心不思議)," "principle (law) wonder (*ri fushigi* 理不思議)"; "the great wonder (*dai fushigi* 大不思議)" belonging to Mahavairocana (*Dainichi nyorai* 大日如来). I believe today's science has identified and has marshalled most of the "substance wonder" (People call this reason (*riyū* 理由) or a principle (*genri* 原理)). However, in reality, it is not principle. It has just dissected the wonder and understood them as phenomena.[21] (18 July 1903) (Minakata 1971: 364–365)

We can see that he elaborated on his idea of "substance," "heart-mind," and "event" which he had described in his letter written in 1894 into something that represents the "wonder" of this world. He also points out that starting from understanding "substance wonder," today's science has identified the "reasons" or "principle" behind the phenomena belonging to "substance wonder". However, according to Kumagusu, what has been identified as the "law" or "principle" of phenomena is only touching upon the part of all wonders. He explains that though "event wonder" has been investigated through mathematics and logic; "heart-mind wonder" through psychology; since they all start from "substance wonder" as the object of the research, science has yet to find the right method to directly investigate "heart-mind wonder" and "event wonder" (Minakata 1971: 364).[22] Following his explanation of wonders and his criticism of science, he starts his explanation of his mandala as follows.

> Beyond "substance", "heart-mind" and "event", there is "principle (law) wonder"... It is curious that in this universe, just as it can be that "heaven" (*ten* 天) is "principle (law) (*ri* 理)", ("principle" is the "line of logic (*sujimichi* すじみち)"), as you can see in my drawing (the drawing can only be drawn on a plane surface. However, it actually should be recognized as three dimensional with length, width, and depth), the "lines of logic (principle)" comes from all directions, from front, back, left, right, top, and bottom, and yet it follows the "laws and principles (*jiri* 事理)" to compose this universe. Such lines are innumerable. Therefore, any chosen line, when it is fully pursued (investigated), could identify everything

[20] *Fushigi* could also be translated as "mystery." I chose to use "wonder" in want of highlighting Kumagusu's appreciation and notion of the "wonderful nature."

[21] See Quotation 4 for Japanese.

[22] In his letter dated 5 February 1894, Kumagusu explains logic and mathematics as being "abstract science;" astronomy, physics and chemistry as being "abstract concrete science;" and physiology, biology, psychology, science, and humanities as being "concrete science" (Minakata 1971: 197).

> and could enable everything... Because of the "*Suiten* 萃点" (literally, an intersectional or convergent (萃) point (点)) of all principles, various "reasons and principles (*shojiri* 諸事理)" can be (more) quickly and easily understood when considered from its perspective.[23] (18 July 1903) (Minakata 1971: 365)

The explanation continues, and he describes how each point in Japanese alphabetical order (*I* イ, *Ro* ロ, *Ha* ハ, *Ni* ニ, etc.) is interrelated. According to Kumagusu, the "lines of logic" in Fig. 5.5 represent the causality of "events" and "reasons and principles", and each point shows the phenomenon which is acknowledged by humans. The Point *I* イ in the center, which is named as the *suiten*, is a trigger-like-point (or a phenomenon) which reveals various principles related to that particular phenomenon. According to TSURUMI Kazuko, the uniqueness of the "*suiten*" is that Kumagusu is not referring to the "*suiten*" as the center of all causalities. Instead, Tsurumi thinks that since the mandala has its own dynamism, the "*suiten*" constantly changes accordingly to the mandala's dynamic condition (Tsurumi 2001: 153). The Point *Ha* ハ in the upper left can be easily detected since it is on the intersection of the two lines of logic, which can be approached from two aspects for analysis. However, since points such as Point *He* へ and Point *To* ト are far from the humans' sphere of recognition and are not intensively connected with other points, it becomes difficult for them to be cognized by humans. Lines such as *Nu* ヌ and *Ru* ル are much farther away, indicating something beyond the current limits of human's knowledge and the human capacity to understand precisely. Humans can suspect the existence of Line *Nu* ヌ since it is slightly attached to Point *O* オ and Point *Wa* ワ. However, Line *Ru* ル's existence above the Line *Nu* ヌ can only be assumed by humans, and it represents the "principle (law) wonder." According to Kumagusu, the wonder outside his drawing represents the "enormous wonder," the Mahavairocana itself: Mahavairocana for Kumagusu is the "strange and superb wonder (*myōfukashigi ga sunawachi Dainichi nyorai nari* 妙不可思議がすなわち大日如来なり)" (Minakata 1971: 364–366).

For Kumagusu, it seems that there is a "knowable principle (*kachi no ri* 可知の理)" (Minakata 1971: 366) and "unknowable knowledge (*fukachi* 不可知)" (Minakata 2010: 277). "Knowable principle" could be known and understood by humans. However, because of the humans' limitation in understanding unfolding phenomenon, Kumagusu thinks that there is "unknowable knowledge" in the universe. Once the knowledge is known, it can be sublimated to principle, however, unless the knowledge is known, the knowledge itself remains unknowable knowledge unidentified by the humans. *Minakata Mandala op.1* could be understood as a drawing to articulate his philosophy and the relations between "knowable principle" and "unknowable knowledge".

In a letter to Hōryu dated 4 March 1894, Kumagusu mentioned that he had been exchanging letters with D. T. SUZUKI (SUZUKI Daisetsu 鈴木大拙) on "ineffable words (or words unable to say/*fukagen* 不可言)" that regards the different notions

[23] See Quotation 5 for Japanese. Though I describe the literal translation of *suiten* in parentheses, there are various interpretations on *suiten* among Japanese scholars. Further research as to defining what really Kumagusu meant by *suiten* has to be conducted.

of the religion and the moral (Minakata 1971: 314).[24] It can be assumed that ever since exchanging letters with D. T. Suzuki on "ineffable words", he had been thinking about the "knowable principle" and "unknowable knowledge" through the scientific approach in understanding the wonder of the universe. To find and to prove scientific laws, scientists have to set up a certain research object and its domain, which is placed in a certain field of research interest. Science can only identify the law or principle based on a set situation with "known" or "identified" sets of items to be observed and recorded. All the "unknown," "unidentified" or "undetected" items are dealt with by setting parameters for adjustment, absorbed in the "known" or "identified" sets of items. For example from contemporary science, the most famous parameter adjustment in global climate modelling concerns the "unknown" reaction of the massive water, the ocean. The absorption rate of the carbon dioxide into the ocean, as well as the heat absorption rate of the ocean, and its feedback mechanism on the global climate is yet unknown in detail. Although it is assumed that the ocean is functioning as a mitigating substance for a rapid climate change, it is not known to what extent the ocean is functioning for climate change mitigation. Therefore, in the climate modelling, the function of the ocean is adjusted by changing the weighting rate of the parameters (For further details see Murphy et al. 2004).

2.3 Concept of "Causality (inga 因果)" and "Dependent co-arising (engi 縁起)": Minakata Mandala op.2

Minakata Mandala op.2 appears in Kumagusu's letter written on 8 August 1903. In this mandala, Kumagusu adds his new concepts of "name (*myō* 名)" and "symbol (*in* 印)". Followed by his explanation on the mandala, he explains "causality (*inga* 因果)" and "dependent co-arising (*engi* 縁起)". Although Kumagusu never mentions *Minakata Mandala op.1* as his mandala, interestingly, he describes that he drew a mandala when explaining *Minakata Mandala op.2* in his letter to Hōryu (Figs. 5.6, 5.7 and 5.8).

In the *Shingon* school, according to Kūkai's teachings, the Womb Realm (*Taizōkai* 胎蔵界) is known as "the world of law (or principle) (*rikai* 理界)" and the Diamond Realm (*Kongōkai* 金剛界) is known as "the world of wisdom (*chikai* 智界)". The Womb Realm mandala (*Taizōkai mandala* 胎蔵界曼荼羅) and the Diamond Realm mandala (*Kongōkai mandala* 金剛界曼荼羅) depict the reality of the Womb Realm and the Diamond Realm, respectively. These two mandalas are always displayed in a set as two sides of the same single body (*nini-funi* 二而不二 or 而二不二 of the *ryōbu-funi* 両部不二, non-duality of two realms, concept of the world) and are considered as an expression of the entirety of Dharma (Masaki 2012: 181).[25] The cosmic view of Kūkai 空海 (774–835) manifests through the mandalas

[24] Although Kumagusu mentions his letter to D.T. Suzuki the original letters are still missing.

[25] Kumagusu uses the Chinese character 二而不二 for *nini-funi*. However, *Shingon* school uses the character 而二不二 when explaining the *ryōbu-funi* concept in general. *Ryōbu-funi* is the world view that understands two identical things as not separated from each other but one single thing.

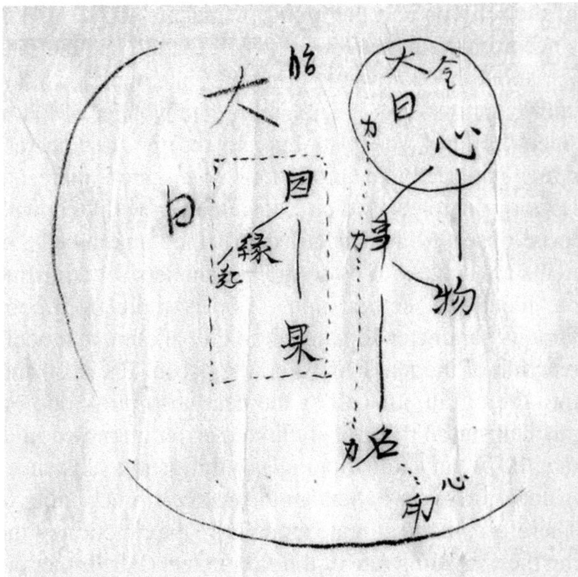

Fig. 5.6 Minakata Mandala op.2 (a). (Copyright authorized by Minakata Kumagusu Archives (Minakata Kumagusu Kenshōkan))

that even though the world is divided in two, both realms are ultimately the same realm, that the "the world of law" and "the world of wisdom" always are understood as a unified one. Kumagusu apparently knew this unity of the two realms; therefore, the uniqueness of *Minakata Mandala op.2* is that Kumagusu combines the Womb Realm and the Diamond Realm into one mandala.

Kumagusu explains his *Minakata Mandala op.2* as follows:

> [T]he Diamond Realm is embedded in the Womb Realm. A part of the "heart-mind" creates "substance" with the nugatory force of the Mahavairocana's "heart-mind" (the part of the Diamond Realm mandala which remains when the "heart-mind" is excluded). An "event" occurs based on the mutual reflective movement of "substance" and "heart-mind". The "event" is transmitted as "name (*myō* 名)", as well as a reaction to such force... "heart-mind", "substance", "name", and "incident" form in various orders, as a reaction to such force... As the drawing shows, the "name" and "symbol (*in* 印)" of the "mantras (*Shingon* 真言)" are not the "appellations (*na* 名)" of "substances". A disappearing "event" (the event is different from "substance" and "heart-mind," thus disappears when it is stopped), remains as "name" in the Womb Realm mandala. When it reflects itself in the "heart-mind," it appears as "symbol." Therefore, creed, language, habit, heredity, and tradition (legend) which are difficult to understand through the philosophy of science, can be proven to "exist (*jituzai* 実在)" in *Shingon* 真言 (the true word or mantra Buddhism), which is the "name."[26] (8 August 1903) (Minakata 1971: 390)

It is clear that Kumagusu elaborates and develops his *Minakata Mandala op.2* from his initial drawing of the endless chain of a causal effect between "heart-mind", "substance", and "event" (see Figs. 5.1 and 5.2). Through the drawing of mandalas, Kumagusu seems to deepen his understanding of the world. The newly

[26] See Quotation 6 for Japanese.

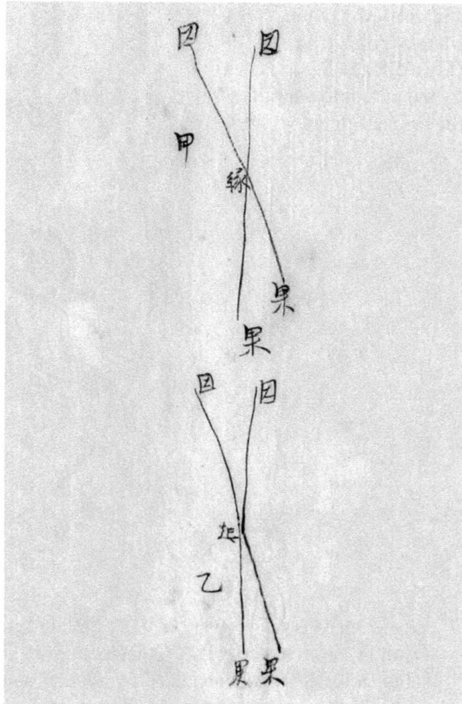

Fig. 5.7 Causality and Dependent Co-arising (a). (Copyright authorized by Minakata Kumagusu Archives (Minakata Kumagusu Kenshōkan))

added idea is shown in his mandala as a whole endless process which starts from the "heart-mind" embedded in the Diamond Realm, with "power (*chikara* 力)" as a driving force flowing from the "heart-mind" to the "symbol" via "event" and "name." Though we can still see the interdependent relationship of "heart-mind," "substance," and "event," "name" and the succeeding "symbol" only appears by-passing the "event." It seems that for Kumagusu, what he called a "name" is a disappearing "event" which remains in the Womb Realm; and the "symbol" is a reflection of the "name" in the "heart-mind" (he shows this idea by connecting a line between "event" and "name" with "effect" in the Diamond Realm). Kumagusu defines "name" as something that represents humanity's discourses and cultures, such as creed, language, habit, heredity, and tradition.

A unique interpretation of Kumagusu is that he explicitly writes that the Diamond Realm is embedded in the Womb Realm. Traditionally, in the *Shingon* school, the Diamond Realm and the Womb Realm are divided into two realms, although it is thought as the two sides of the same universe. According to Kumagusu, the Womb Realm is represented by "causality" and "dependent co-arising" which is the law of the universe. Following his explanation of *Minakata Mandala op.2*, by showing Fig. 5.7 (or Fig. 5.9), Kumagusu explains the difference between "causality" and "dependent co-arising" as follows:

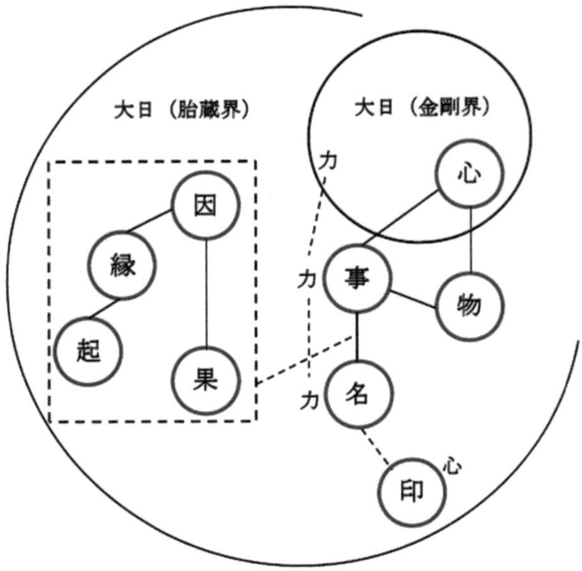

Fig. 5.8 Minakata Mandala op.2 (b). (Figs. 5.8 and 5.9 is drawn by the author to simplify the original figures)

> Where there is no "cause (*in* 因)", "effect (*ga* 果)" does not happen. Different "causes" result in different "effects." "Coincidence (*en* 縁)"[27] occurs when one continuation of "causality (*inga* 因果)" is interfered by another continuation of "causality" (as shown in figure *kō* 甲, Kumagusu meets an elementary school teacher on a mountain). When the interference occurs and has some impact on the former continuation of "causality", "origination, creation or arising (*gi* 起)" occurs (as shown in figure *otsu* 乙, Kumagusu talking with that teacher and the next day, decides to pay a visit to an old former teacher of that teacher). Therefore, we, in our own body, are experiencing the continuation of various "causalities" at this very moment. As for "coincidence," we meet countless numbers of "coincidences" in the blink of an eye (at every instant). These facts, how we control our "heart-mind" and "body (*karada* 体)" "originates, creates, or arises (*gi* 起)" an "event," and together with the continuation of the "causality effect" of our "actions (*kōdō* 行動)," the "track or path (*kidō* 軌道)" deviates or the deviated "substance" returns to its own "track (path)." [Thus] the conclusion is drawn from my Mandala; "it is said, (we should) say things are not a burden," should be the answer and my argument stops here.[28] (8 August 1903) (Minakata 1971: 391)

Here, Kumagusu divides the "causality" and "dependent co-arising" into individual Chinese characters and explains the difference between "coincidence (*en* 縁)" and "origination, creation or arising (*gi* 起)" in relation to "cause (*in* 因)" and "effect (*ga* 果)". In Fig. 5.7, to differentiate between his idea of "coincidence (*en* 縁)" drawn in figure *kō* 甲 (upper of Fig. 5.7, left side of Fig. 5.9) and "origination, creation or arising" drawn in figure *otsu* 乙, Kumagusu has drawn the line of "effect"

---

[27] Here I chose to use "coincidence" to translate the word *en* 縁. In Japanese, *en* has several connoted meaning as chance, karma, relationship, bond, fate, and destiny. I chose to translate *engi* 縁起 as "dependent co-arising", by considering the Japanese context of the word *en*. However, I think here, *en* is better translated as "coincidence" as to emphasize the striking occurrence of two or more events at one time (or instantly) happening by mere chance, like a coincidental event.

[28] See Quotation 7 for Japanese.

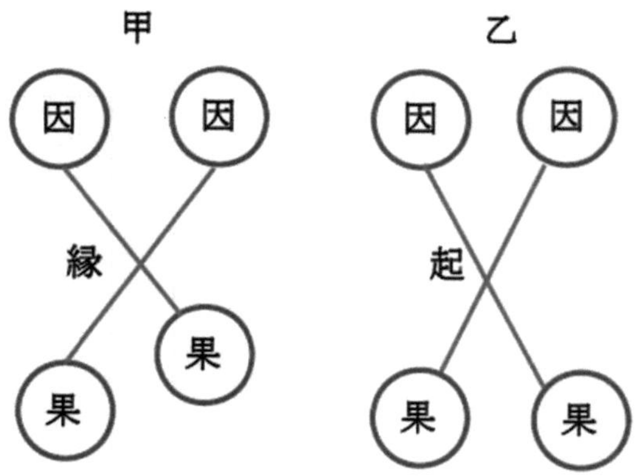

Fig. 5.9 Causality and Dependent Co-arising (b)

with a slightly different angle and length. That the "track (path)" of "causality" changes accordingly to "coincidence" and whether the "coincidence" stays as a "co-event" or "coincidence" "originates, creates, or arises" is the difference between a "coincidence" which stops there as a "co-event" and a "coincidence" "originating, creating, or arising" a new "event."

Figure 5.10 shows the mandala style conversion which I have constructed based on Kumagusu's explanation of *Minakata Mandala op.2*. In the Diamond Realm, which is in a three by three matrix format, five wisdom Buddhas are placed in the order as shown in Fig. 5.11, starting from *Jōjin-e*成身会 numbered as ①.[29] Therefore, I put the "heart-mind 心" in the centre of the Diamond realm, surrounded by "substance 物," "event 事," "name 名," and "symbol 印," following the mandala custom and Kumagusu's explanation.[30] Here, the "heart-mind" represents the function of the universe that reveals itself before humans through various forms; as in the form of "substance," "event," "name," and "symbol." As for the Womb Realm, "causality (*inga* 因果)" and "dependent co-arising (*engi* 縁起)" are placed outside the Diamond Realm but within the Womb Realm. The diagonal relationship of "cause" and "effect" is the piercing law that rules the Diamond Realm, constantly affected by the continuous flow of "power as a driving force (*chikara* 力)" from the "heart-mind." I believe Kumagusu is thinking of a somewhat layered structure of the Diamond Realm, the Womb Realm, and the Universe. Though I have omitted the Mahavairocana in Fig. 5.10, since Kumagusu describes Mahavairocana as being in both the Diamond and Womb Realm (Fig. 5.8), it is likely that, in the very centre of

[29] The Diamond Realm mandala is drawn in a three by three matrix format, representing the nine stages of Buddha. Therefore, Diamond Realm mandala is also known as Nine-stage mandala (*Kue mandala* 九会曼荼羅). Though it is numbered for convenience, in Esoteric Buddhism there is no hierarchy among the five wisdoms.

[30] Kūkai, in his book "*Han'nya Shingyō Hiken* 般若心経秘鍵", explains his understanding of "Heart Sutra, *Han'nya Shingyō* 般若心経". The book explains that "heart (*kokoro* 心)" stands for the "True Word (*shingon* 真言)," the very essence of his philosophy (Takemura 2015: 132). Therefore, Kuamgusu's idea of setting the "heart-mind' in the centre of his mandala is compatible with Kūkai's philosophy.

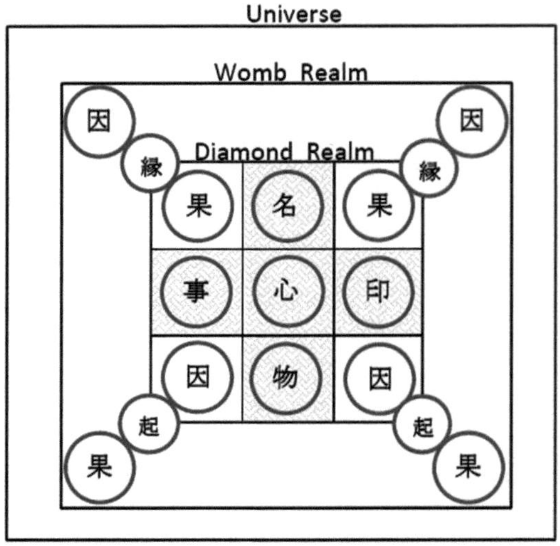

Fig. 5.10 Mandala Representation of Causality and Dependent Co-arising

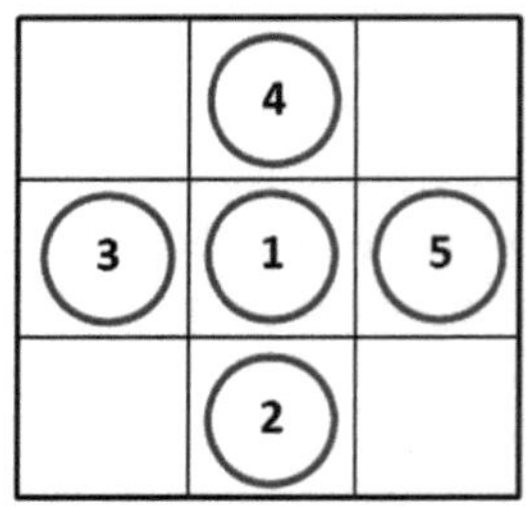

Fig. 5.11 The Order of Five Wisdom Buddhas

the layered structure of the realms, Mahavairocana is vertically piercing all the layers, and is considered as the substance providing the original force of the "heart-mind."

As Kumagusu writes the word, *nini-funi* (二而不二 or 而二不二) in his letter of 8 August 1903, it can be presumed that he is well aware of the characteristics of the mandala, that the Diamond Realm and the Womb Realm represent the two sides (aspects) of the same one body. I read Kumagusu's combining of the two realms as his attempt to express the layered structure of continuous flow and the chain reaction of the unfolding and developing interrelations among "heart-mind," "substance," "event," "name," and "symbol" which has no beginning and no end. And with his attempt to combine the two realms, I think, he tried to prove that what had been divided into two realms, the world of action to gain wisdom (represented in the Diamond Realm) and the world of principle where law rules (represented in the Womb Realm) could be neatly combined into one single realm, symbolising his ideology on science.[31]

[31] Although I am not going to explore in detail, because of the limitation of the space, the concept of *nini-funi* may also be applied to the ideal usage of science and technology: Womb Realm represents science and Diamond Realm represents technology. Without careful interplay and mutual

What is interesting about his idea is that by including the concepts of "name" and "symbol," he tries to include humanities (liberal art) studies in the field of science, suggesting that the liberal art studies are not different from science but are intertwined together with his concept of "causality and "dependent co-arising." Furthermore, through his thought exercise to explain the interrelations of phenomena, he develops his initial idea of the triangular relationship of "heart-mind," "substance" and "event" into a very complex system of the universe. With this mandala, Kumagusu challenges the limitation of science from the aspect of its simplicity in analysing only the limited dimensions of the fact that humans cognize. For Kumagusu, science needs to develop further to be able to analyse the intertwined, inter-dependent, and co-arising relations that exist beyond humans' cognition. This leads us to speculate on Kumagusu's ethics regarding the use of science and technology in our modern society. In other words, Kumagusu's ethics could be understood as practical ethics that unfolds between an intersection of applied science and liberal art studies.

Here, I would like to summarize what was covered in Section 5.2 in terms of Kumagusu's critical view of modern science and his philosophical thought-experiments or his attempt to solve the problem embedded in science.

1. He criticized science for focusing mainly on the phenomenon of "substance." By dividing the universe into three aspects; "heart-mind," "substance," and "event" he tried to explain the mutual interaction and reaction based on "causality" occurring between "heart-mind" and "substance" which could be categorized as "event."
2. He indirectly criticized the linear concept of time embedded in science, the past following the present, and the present leading to the future. By explaining his chronological concept based on *Avatamsaka Sutra*, a time which has no beginning and no end, which spreads, unfolds and interacts like lace-work, between the present, past, and future, he suggested a different concept of time that could be applied to science.
3. He criticized science for developing only through "substance wonder" and claimed that it needs to focus on "heart-mind wonder" and "event wonder." By showing *Minakata Mandala op.1*, he explained that there are "knowable" and "unknowable" principles, that humans can only identify the "unknowable" through the "knowable." And he pointed out that there is a possibility that humans are incapable of knowing the "unknowable."
4. By showing the *Minakata Mandala op.2*, he explained his new concepts of "name" and "symbol" that represent the humanities and argued that science is not limited to "science," but is open to and linked with the humanities. He also made an attempt to combine *action (wisdom)* and *principle (law)* under the law of "causality" and "dependent co-arising." Through this, he was trying to express the ambiguity of the universe, that nothing is ever stable, but always in movement and in the process under the law of "causality" and "dependent co-arising."

complements, both science and technology could lead the society into unintended outcomes as observed in the 2011 Fukushima incident.

3 Kumagusu's Ethical Implications: Choices in the Use of Technology

Through his understanding of the limitations of science, as described in Sect. 5.2, in his later life Kumagusu initiated an action against the Shrine Merger Order as is briefly mentioned in Sect. 5.1. The purpose of his activity was to preserve the ecology of the shrine forests, and he argued that once a particular ecology is destructed, it can never be restored as it existed. He is known as the first Japanese to introduce the word "ecology" to Japanese (Minakata 1971: 526) through his activity. I would like to expand his argument on science to the environmental problem we face today, in link with the use of technology, to consider what could be ethical implications that could be drawn out from Kumagusu.

3.1 Climate Change Science and the Tentative Technological Solution

In order to tackle the global climate change issues, nuclear power has been recognized as one of the least carbon-emitting, cleaner energy options to generate electricity during its electricity generation process (IPCC TAR 2001). In the last five reports published by IPCC (Intergovernmental Panel on Climate Change), scientists involved in the process showed a shifting, unstable position towards nuclear power (World Nuclear Association 2015). However, in its most recent Fifth Assessment Report, published in 2014 (AR5), IPCC scientists grouped nuclear power as if it was in a renewable energy category (IPCC AR5 2014). In this section, (1) by touching upon the current trend of global society trying to introduce or implement large-scale technology, particularly nuclear power plants and geoengineering technologies, in order to mitigate or to adapt to climate change and (2) by following Kumagusu's views on science, I would like to point out the potential problems embedded in introducing such large-scale technology without fully knowing, understanding or considering beforehand the possible consequences of installing those technologies. What could be a probable ethical reaction we can take towards uncertainties and risks related to large-scale technology?

In our current society, we are now facing various global environmental problems, such as acid rain, ozone depletion, climate change, and biodiversity losses. All these problems are analysed as having certain anthropogenic reasons as their cause. The once local environmental problems of the 1960s gradually expanded to trans-border environmental problems in the 1970s, exemplified by the acid rain, which severely affected the Scandinavian countries' forest and lake ecosystems. For problems such as acid rain and ozone depletion, since the sources of the problems were comparatively easy to identify within the scientific research format, compared to systematically understand the impacts of the identified causes of the problems, the measures proposed for problem-solving were to simply to decrease or to control emission levels to a manageable level of the identified "cause" or to ban the use of the identified "cause" itself. In the field of environmental protection, new ways of approaching uncertainties such as the Precautionary Principle was introduced in 1992 during the Rio Earth Summit to avoid serious or irreversible damage (Rio Declaration on Environmental and Development 1992: Principle 15).

In the case of decreasing or controlling the emission level with respect to the acid rain issue, with regulations introduced at the national level, sulphur dioxides, and nitrogen oxides emission levels were decreased by introducing, amongst others, end-of-pipe technology.[32] On the other hand, in the case of climate change, GHGs (Green House Gases) identified as six gases, such as carbon dioxide and methane have been measured at the national level, compiled under the UNFCCC (United Nations Framework Convention on Climate Change) and within the ratified countries the attempt has been made to decrease their emission to a controllable level under the Kyoto Protocol since 2005.[33] In December 2015, the Paris Agreement was proposed after the COP 21 which stresses the importance of collective action of all nations for the next generations (UNFCCC Paris Agreement 2015). In the case of banning the identified "cause" itself, the Montreal Protocol was ratified in 1989 by 20 countries that agreed to gradually ban the use of ozone-depleting substances, such as HFCs (hydrofluorocarbons) by 1996 for developed countries and by 2015 for developing countries.

However, the question remains: can we trust our modern science to capture what is really happening? As Kumagusu has argued, modern science is based on "causality" identifying the "causal relationship" of the cognized "substance" through certain phenomena. Is this scientific attitude really covering all "causalities"? Are the measures identified through scientific studies feasible enough to tackle the problems realised? Humans can realise and cognize what could be perceived as information only through their limited view as being humans. In that sense, the cognition of time scale and the spatial scale is always limited to the human perception of time and space. Though science is catching up, humans still are having difficulty and are at a stage of progress in knowing and understanding a causal relationship happening at a low-frequency rate in a very long time span of macro-cosmos as well as the causal relationship happening at a high-frequency rate in a nano-cosmos. In this regard, humans have not well yet grasped the causal relationship happening between the nano-world and the macro-world.

Kumagusu's arguments questioned this very notion of anthropocentric recognition and analysis of natural phenomena: anthropocentric cognition could be problematic in a way for humans not being able to realise nor recognize the possible unidentified "causation" which is caused by limitations in human cognitive capacity. This undetected causality is experienced by humans as unpredictable butterfly

---

[32] End-of-pipe technology is a technology and a method used to remove formed contaminants in the last stage of a process before air, water and soil are disposed from chemical processes.

[33] There are six gases identified as GHGs under Kyoto Protocol namely; carbon dioxide (CO2), methane (CH4), Nitrous oxide (N2O), Hydrofluorocarbons (HFCs), Perfluorocarbons (PFCs), and Sulphur hexafluoride (SF6).

effects or boomerang effects, from time to time. Science is, after all, the analyzation of "causal" relationship which happened in the past, on what had been identified as a phenomenon based on human perception. In other words, there should be unknown interconnectedness that is not captured by the human radar, in other words, hidden "causal" relationships between the revealed consequences.

Measures that were taken in response to scientific findings were either "passive" or "proactive" reactions after the identification of "causality". Because of the limitations of technological development and science in identifying the "causal effect," humans are only reacting to the issues that are perceptible enough to be cognized as problems. However, with the improvements in technology and a further understanding of "causal relationships" and its "effects," in the twenty-first century, our current society has gained the ability to predict the future to a certain level and to take action before the predicted severe damage takes place. It is still fresh in people's memory that more than 16,000 people evacuated from three different villages surrounding Mt. Usu in fear of volcanic eruption in Hokkaido in 2000, based on the prediction made by a scientist (Volcano Science Museum 2000). However, simultaneously, there are various natural disasters which occurred without being predicted by the scientists. Humans have also gained technology to ease the "causality effect" through efforts by inventing new technologies. To give a solution to the dilemma of generating a large amount of electricity and to tackle global climate change issues, humans have in our hands a set of technologies to intentionally manipulate the nuclear fusion reactions or to manipulate climate on a large scale, known as geoengineering technologies (IPCC Expert Meeting on Geoengineering 2011), for the benefit of the human society.[34] Both of the technologies have their own pros and cons in usage. The benefit can be large if and when operated in a manageable way. However, the disaster could also be large if we fail to control the use of such technology. Therefore, we now face choices: a choice to take the usual "passive" reaction or to take "precautionary" measures (prevention) before the severe problem occurs.

[34] According to the IPCC, geoengineering technologies are defined as follows: a broad set of technologies that aim to deliberately alter the climate system in order to alleviate the impact of climate change. Most, but not all, methods seek to either (a) reduce the amount of absorbed solar energy in the climate system (Solar Radiation Management) or (b) increase net carbon sinks from the atmosphere at a scale sufficiently large to alter climate (Carbon Dioxide Removal). Solar Radiation Management (SRM) refers to the intentional modification of the Earth's shortwave radiative budget with the aim to reduce climate change according to a given metric (e.g., surface temperature, precipitation, regional impacts, etc.) The well-known two examples of SRM techniques are artificial injection of stratospheric aerosols and cloud brightening. CDR methods refer to a set of techniques that aim to remove CO_2 directly from the atmosphere by either (1) increasing natural sinks for carbon or (2) using chemical engineering to remove the CO_2, with the intent of reducing the atmospheric CO_2 concentration. CDR methods involve the ocean, land, and technical systems, including such methods as iron fertilization, large-scale afforestation, and direct capture of CO_2 from the atmosphere using engineered chemical means.

3.2 *Myth and Magic of Numbers: Uncertainties Reproduced*

As I have argued, Kumagusu developed his own ideal science through his attempt to describe the "event wonder." Here, I would like to expand on what has been argued through Kumagusu's view of science and would like to draw out what could be considered the ethical implication of *Minakata Mandala*, addressing implementation of large-scale technologies, such as nuclear energy and geoengineering technology. In this attempt, I will again quote some of his words from his correspondence with Hōryu.

In his letter dated 8 August 1903, Kumagusu criticized the over-estimation and excessive emphasis on the importance of "numbers" in science. He is apprehensive of the magic of "numbers," that once the "events and substances" are interpreted and symbolised into "numbers," humans tend to recognize the originally ambiguous "events and substances" as something attributable to a fixed or stabilized object:

> Science today depends too much on "numbers (*kazu* 数)" as if they are unquestionable. However, "numbers" themselves are representing the unfixed (indefinite, indeterminate, changing) substances... "Numbers," therefore, are not as definite as scientists think. We measure the slightest change in the "events and substances (*jibutu* 事物)" of the universe with "numbers." However, "numbers" actually exist only through those changing "events and substances"... "Numbers" never represent the essence of the "events and substances," only representing the "characteristics (quantity) (*seisitu* 性質, here Kumagusu uses the word "qualify")" from the outside... It is too hasty to say that things which cannot be defined and represented by "numbers" do not exist. Today's science ignores things which cannot be measured by "numbers."... For these reasons, we should know that today's science is inadequate (unsatisfying)... Scientists say that everything can be understood through science and talk as though things which cannot be understood through science are nothing but a lie. This is same as saying that the world is a machine, humans are animals, and morals and so forth are all falsehoods.[35] (8 August 1903) (Minakata 1971: 393–395)

Though nuclear energy has been employed since the 1950s, the use of the technology is veiled with numerous uncertainties. With regards to the uncertainties of the technology, the implementation of geoengineering technology is still in the early stage of its development. Though the geoengineering technology itself has been known since 1965, when the Report of the US President's Science Advisory Council was issued, none of the proposed methods assessed so far offers an immediate solution to climate change (Keith 2000). Both nuclear power generation and geoengineering technologies gather attention with high expectations as a major technology to tackle climate change. However, further investigation is needed since little is understood of their potential future risks and uncertainties. Neither their positive nor negative effect in the shorter or longer term, nor their substantial and unintended impact to the ecosystem is fully understood. Though there have been intensive arguments among specialists, the scientific evidence is still too weak to justify their implementation for the sake of environmental protection, which leaves us a room to pose a question for ethical consideration on whether humans should utilize those technologies or not.

---

[35] See Quotation 8 for Japanese.

As Kumagusu continuously argued in his letter to Hōryu, current scientific methodology attributed is based on identifying the "causality" of the human "perceptible (cognisable)" things. In Kumagusu's word, "substance" is always the gateway to understanding the logic behind the phenomenon we observe. This implies that, though humans can cognize and analyse the "causality" of the perceived or sensed "substance," humans can never see the universe beyond human recognition. When "substance" or research objects are shown in a format of "numbers" as quantitative amounts, it is easy for humans to mistakenly believe that these objects are well understood and are manageable under human control.[36] Kumagusu explicitly points out the danger of invoking the "myth" of "numbers," in other words, the misperception of "substance" being "under human control" easily occurs when objects are presented in numerical form. He argues that quantified data are just representing a certain aspect (part) of "substance" and may not be representing the ambiguous "substance" itself, which is subject to change within the law of "causal relationship" and "dependent co-arising." His argument can easily be applied to the problems embedded in large-scale technologies: 1) when technology becomes larger, it is obvious that the numbers of the related "substances" will rise, which will lead to further difficulty in grasping the "causal relationships" and "dependent co-arising" force between "substances" and "events"; 2) the quantified data of the identified "substances" and "events" itself might be the source of the ambiguity of the technology which leads to the uncertainties and risks relating to the technology itself. The argument made by Kumagusu indicates that the risks or uncertainties rooted in science and technology can never be removed because of the limited cognition based on humans' ability.

Kumagusu also points out the importance of identifying the "coincidences" and not to focus only on "causality." He further goes on to claim that science needs to go beyond "causality" to find the "meta-causality" covering all phenomena:

> [T]oday's science can grasp (or holds the possibility of grasping) "causality," but can never grasp "coincidence (*en* 縁)." Our mission (task) is to study "coincidences". "Coincidence" occurs when "causalities" are intricately mixed, therefore, our mission is to search for the "encompassing causality (meta-causality)," a layer above all causalities" (*shoinga sōtai no issō ue no inga* 諸因果総体の一層上の因果).[37] (8 August 1903) (Minakata 1971: 391–392)

When recalling Kumagusu's argument on "coincidence" and "dependent co-arising," he argues that the track (path) of "causality" changes with the effect of "coincidence." In Kumagusu's words, either to leave the "causality" "track (path)" as it is, or to change the "causality" "track" through "dependent co-arising". Humans cannot predict which "track" is going to be taken beforehand; therefore, humans can only detect "causality" through the *probability* of occurrence. With this thought in mind, we can easily understand that the partial optimization of the detected "causality" may not be equal to the overall optimization of the whole, including the

[36] Though I acknowledge that there are various arguments on the relationship between "quality" and "quantity,", in this paper, I focus on the problem embedded in "quantifying".

[37] See Quotation 9 for Japanese.

surrounding environment. Even if we may be able to secure partial optimization, with the possibility of "dependent co-arising" occurring unbeknownst to humans (because of its low occurrence (or frequency) rate), this one occurrence of "dependent co-arising," may lead to the corruption of the whole ecological system. This is known today as the "black swan theory" that describes an event that happens surprisingly beyond humans' empirical knowledge with massive impact and is given rational reasons only with the gained hindsight, often using Bayesian statistics for its explanation (Taleb 2007).[38]

Taleb proposes the idea of "metaprobability (or metadistribution)" to assess the degrees of uncertainty regarding risks using statistics (Taleb 2015: 170). Interestingly Kumagusu, without any knowledge of contemporary statistics, proposes a similar notion to Taleb's to find the "meta-causality" which encompasses all the causalities in the universe to avoid disastrous situation caused by human beings. Scientists may see this attempt in finding the "meta-causality" challenging and exciting. However, without any clue to identifying this "meta-causality" ruling our universe, is it really safe to endeavour and implement large-scale technologies in our society full of risks? In other words, with the current scientific capability and scientific methods humans have gained, it is impossible to identify the "meta-causality" that rules our universe. When naively following Kumagusu's view of science, we cannot deny the fact that there may exist a high possibility of "unknowable" consequences taking effect that is "unknown" to us due to our inadequate understanding of the Earth's climate and its relation to outer space.[39] The irreversibility of the "cause" itself (in this particular case, "cause" stands for the implementation of large-scale technologies), along with "unimaginable" natural phenomena (like earthquakes and volcanic eruptions), in resonance with "unthinkable (inconceivable)" human error and mechanical failure, we can never deny the possibility of another "black swan" event occurring anytime soon.[40] All we can do is to build a resilient and robust society anticipating the occurrence of such events. When we cannot deny that there might exist major uncertainties and risks regarding the effects of large-scale technologies affecting the ecosystem, climate system, and other natural systems (including the possible impacts on humans), what could be our possible ethical reaction to using such technologies?

---

[38] "Black swan theory" or "Theory of black swan events" is a metaphor describing unexpected events of large magnitude and consequence that comes as a surprise. Taleb got the original idea from his analysis of financial crises and extended his concept to other historical events such as the rise of the internet, the dissolution of the Soviet Union, and the September 2001 attacks. Since the "black swan" events are impossible to predict, Taleb proposes to build robustness in society in preparation for such huge negative impacts.

[39] Non-naïve way could be to blindly use any possible technologies in hands without considering the consequences of their usage.

[40] After the 11 March 2011 Great East Japan Earthquake, the government officials and representatives of TEPCO (Tokyo Electric Power Company) were repetitively using the word "unimaginable, unthinkable situation (*souteigai no jitai* 想定外の事態)".

3.3 Acknowledging Limitation of Science

Science seems to be touching upon the eternal *truth* and *beauty* of nature and the mind. We would want to think, therefore, that developed technology based on scientific findings embeds the *good* in a sense that helps humanity flourishes when it is used in a proper way.[41] However, we have to be aware of the fact that, even though the initial intention of using the technology was good, and even when the technology had been properly used, with the undetermined "unknown" "causal" and "dependent co-arising" effect, it may have a huge, irreversible impact on human society in the long-term, which we may call as uncertainties and risks related to science and technology implementations. The solid fact of humans being the only organisms able to analyse "substances" in a scientific manner indicates that humans are constrained to the anthropocentric worldview, which limits the humans" decisions and active choices to human perceptions of the world as well. Based on this way of thinking, science – which we consider as being something solid and determined – would be recognized as something invented by and limited to the human cognitive system. By following this line of thinking, the fictional "absoluteness" of science could easily be denied. After all, humans see the world in the way we want to see it and make judgements based on what we have seen and understood. The different worldview deriving from unique interpretation on Buddhism that Kumagusu has shown implies of a continuous net of reflection, interaction, "relational causality" and "dependent co-arising," constantly working in resonance among "substances" without being detected by us, humans. In that worldview, there is no superiority, nor inferiority, nor predominance. Rather, there exists a continuous flow of interrelations complexly intertwined, mirroring each other, just like Indra's jewel net. In that evolving complexity of interwoven and changing processes, how could we apply ethics in the use of large-scale technology?

If there is any ethical implication which could be drawn from Kumagusu's thought on science, it would be seemingly dualistic but ultimately monistic ethics, like the mandala's ideology of *nini-funi*, implicating both sides of the same one body. When we see large-scale technology as a process (or *action*) as in the Diamond Realm, the "unknown" consequences deriving from such large-scale technology could be thought of positively as another "driving force (power)" to generate new "causality" and "dependent co-arising." However, if we see it as logic (or *principle*) as in the Womb Realm, the technology may be taken negatively as an unknown "cause" which has a possibility to reflect on us as risks and dangers. The ethics implied by Kumagusu would not be hindering the implementation of the technology itself. Rather, what could be considered as Kumagusu's ethics would be that the

---

[41] Most advanced technologies originally derive from military-purpose technological development, including geoengineering technology. The argument can easily be made on the pros, cons, and its social impacts of developed technologies and their usage. The examples of such technologies in question are the nuclear power plant (the peaceful use of atomic bomb technology), computer science (including Artificial Intelligence), high-speed telecommunication, internet, molecular biology, and its related stem-cell technology, etc.

humans have to be prepared in taking responsibility for "unknown" consequences, risks, and uncertainties when applying a technology, which may lead us to actively choose not to use such technology in order to avoid the "unimaginable" and "unthinkable" harmful impact that might occur, which has a possibility to cause unintentional disaster through the implementation of the technology. Whichever choice we make, either to implement the technology or not, the key lies in whether we will be ready to take responsibility for the uncertainties and risks embedded in the new technology, seen from a long-term perspective. Furthermore, in that sense, Kumagusu's thought implies an ethics of the necessity for humans to be sensitive to the "unknown," the "unimaginable," and the "unthinkable" in introducing new technology to the universe.[42] A new kind of ethics that hinders humans from blindly using technologies in hand, but requires a second thought on the invisible consequences that might occur before using the technology.

4 Conclusion: Towards an Ethics Leading to a "Peace of Mind"

As argued, Kumagusu challenged the worldview of modern science and pointed out the limitations of science through his thinking based on Buddhist thought. Kumagusu's main intention was not to argue that Buddhism implies a superior universal view. Instead, by utilizing the Buddhist cosmology, his intention was to better science itself and to point out the limitations embedded in the modern scientific view. By following Kumagusu's thought concerning his ideal science, Kumagusu's ethical implication is drawn out: ethics of the necessity for humans to be sensitive to the "unknown," the "unimaginable," and the "unthinkable." His ethical implication may lead us to stop and reconsider the current trend in blindly trusting and implementing technologies, without considering beforehand the "unknown" and "invisible" consequences that might occur.[43]

If a large-scale technology, the intentional manipulation of nature based on our scientific understanding, is to be implemented in the real world, we have to be aware, that humans are simultaneously introducing a new dynamic impact which our Spaceship Earth has never experienced before (Stevenson 1965).[44] There could

[42] The author is fully aware, in terms of ethics; Kumagusu could be straightforwardly dealt with in regard to his opposition against Shrine Merger Order. However, when appreciating his knowledge and argument on science and scientific approach in combination with his unique understandings on Buddhism, I think it is much more important and thought provocative to interpret his earlier works into ethics.

[43] Environmental assessments are encouraged in most of the countries before implementing large-scale technologies. However, environmental assessments are limited to assessing objects that is already recognized as environmental indicator.

[44] The concept of Spaceship Earth is known widely through Adlai Stevenson's (American Ambassador to the United Nations) speech to the UN made in 1965. The analysis on the background to Stevenson's speech in relation to the Cold War is done by Richard Deese. The thinker,

be an argument made based on the Buddhist cosmology, of an impermanent, ever-changing universe, implying the existence of multiple universes (multiverse as is argued in the field of quantum mechanics). However, in our global society, rather than stressing the difference embedded in the religious beliefs of the East and West, humans have to overcome the cosmologies built on such religious beliefs to better the global society itself. In this sense, Kumagusu's argument on the limitation of science and his ethical implication towards science and large-scale technology per se could be one of a starting point for humanity to reconsider the blind usage of science and implementation of large-scale technology.

Kumagusu has already written in 1902 on the "sense of security" or "peace of mind (*anshin* 安心)," by referring to a now famous problem in computer science: "ten divided by three times three" does not equal to ten, but is calculated as 9.9999∞.

> When 10 is divided by 3, the answer will be 3.33333333+ repeated a few billions of times in 3, and yet it wouldn't reach the accurate answer; two 3s are better, and three 3s are far better than two in terms of accuracy. When comparing the multiplication of 3.3 to 3 (9.9) and 3.33 to 3 (9.99), the latter gets one nine-hundredth closer to 10. People in this world like to have a "peace of mind (sense of security, *anshin* 安心)." A "Peace of mind" is the ultimate joy. However, there are degrees of "peace of mind." There exists 9.9 "peace of mind," 9.99 "peace of mind," and 9.999+ "peace of mind." Even though it never reaches 10, the deeper "peace of mind" is closer to 10.[45] (22 March 1902) (Minakata 1971: 316)

From this passage, we can also draw out the ethical implication for managing uncertainties and risks. Because of the problems related to "number" itself, even with a full spec computer with a detailed future forecasting model installed, though we may be able to know the trend through computer modelling prediction, the future will always remain "unknown" to us. In this universe that is full of "mysteries" and "wonders" yet to be unveiled by our science, we are observing the phenomena of the universe through our limited field of anthropocentric vision. Just like looking at the world through microscopes and binoculars, we may be seeing the focused field in detail, and may precisely know and understand what we are observing from humans' viewpoint. However, we have to bear in mind the unfocused area still remains a "mystery" and a "wonder" to us. Therefore, the ethical choice is to make relentless efforts to improve our science to the approximate value of ten as indicated in Kumagusu's word, a "peace of mind," and to constantly reflect on ourselves and our use of science and technology.

Kumagusu writes that the superiority of humans lies in humans having the "heart-mind" to enjoy the universe and to translate it into "wisdom." Science only accounts for part of the "wisdom." "Wisdom" supersedes academic disciplines and is right there in front of our eyes which are working relentlessly even in the smallest phenomenon of our daily life. Such phenomena rely solely on our "heart-mind" to be sensed, cognized and understood. In short, Kumagusu implies that without

Buckminster Fuller, who took up Stevenson's idea is also famous for advocating the term Spaceship Earth in terms of environmental protection. The detailed analysis on Fuller's intention is done by Peder Anker.

[45] See Quotation 10 for Japanese.

careful interplay and mutual complements of science and technology, both science and technology may lead the society into unintended outcomes such as observed in the 2011 Fukushima incident.

> All nature of the universe is infinite. However, humans have "heart-mind." As long as they have "heart-mind," they receive joy from the "universe (*uchū* 宇宙)" to the extent permitted by their "heart-mind's" capacity. They translate (change) the slight part of the universe into joy in their "heart-mind." I think this whole process is called as "wisdom (*chi* 智)."… Today, when all the scholars in the world wish to become philosophers, it (becoming a philosopher) can never be accomplished through reading books and cramming knowledge into their brains. It can only be accomplished through enjoying the infinite phenomena of the universe, even just one slight phenomenon, examined (considered) through the "heart-mind" makes this possible.[46] (30 June 1903) (Minakata 1971: 329–330)

Overall, Kumagusu's ethics suggests both a "weak (passive)" and "strong (active)" ethics, as is indicated in the *nini-funi* concept (the different sides of the same one body) of the Diamond Realm mandala and the Womb Realm mandala. Kumagusu, throughout his correspondences with Hōryu, though he could not propose the actual solution and methodology to better modern science based on "causality," he pointed out the limitations of science and presented his ideological science as including the concept of "dependent co-arising." Therefore, from his thinking, only a "weak ethic" could be implied, which neither accelerates nor hinders the progress of modern science based on "causality" and excessive development of technology. However, at the same time, his ideal science is indicative of a very "strong ethic." It is a very "strong ethic" which requires humans to take responsibility for the consequences of the implementation of such technology; to take into consideration in advance the "unknown," "unimaginable" and "unthinkable" ripple effect, boomerang effect, and butterfly effect which may affect our Spaceship Earth. A "strong ethic" that requires humans to exist in resonance with other "heart-minds" and "substances" of the Universe. Regarding our use of science and implementation of large-scale technologies, Kumagusu's "strong ethic," which requires humans to be more thoughtful of the hidden consequences of our doing toward nature, might well fit into our contemporary society where restoring our "peace of mind" is an imperative need.

Quotations

Quotation 1

小生の事の学というは、心界と物界が相接して、日常あらわる事という事も右の夢のごとく[…]。心界が物界と雑りて初めて生ずるはたらきなり。[…]　事にはそれぞれ因果のあることと知らる。(明治26年12月21日、河出文庫62、平凡社7巻145)

[46] See Quotation 11 for Japanese.

Quotation 2

今の学者(科学者および欧州の哲学者の一大部分)、ただ箇々のこの心この物について論究するばかりなり。小生は何とぞ心と物とがまじわりて生ずる事(人界の現象と見て可なり)によりて究め、心界と物界とはいかにして相異に、いかにして相同じきところあるかを知りたきなり。(明治26年12月21日、河出文庫63、平凡社7巻146)

Quotation 3

上の図は網の目のごとく、二集まって一となり、一散じて二となるように、二倍ずつのものとせるが、実はこれどころのことではなく、下の図のごとく、レースをあんだように、百集まりて一となり、また分かれて百となるようなものと見れば、大いによく分かるなり。(明治27年3月3日、文庫p146–147、平凡社7巻 210–212)

Quotation 4

不思議ということあり。事不思議あり。物不思議あり。心不思議あり。理不思議あり。大日如来の大不思議あり。予は、今日の科学は物不思議をばあらかた片付け、その順序だけざっと立てならべ得たることと思う。(人は理由とか原理とかいう。しかし実際は原理にあらず。不思議を解剖して現像団とせしまでなり。[...])(明治36年7月18日、文庫295–297、平凡社7巻364–365)

Quotation 5

さて物心事の上に理不思議がある。[...]　さて妙なことは、この世間宇宙は、天は理なりといえるごとく(理はすじみち)、図のごとく(図は平面にしか画きえず。実は長、幅の外に、厚さもある立体のものと見よ)、前後左右上下、いずれの方よりも事理が透徹して、この宇宙を成す。その数無尽なり。故にどこ一つとりても、それを敷衍追求するときは、いかなることをも見出だし、いかなることをもなしうるようになっておる。[...] 諸事理の萃点ゆえ、それをとると、いろいろの理を見出すに易くしてはやい。(明治36年7月18日、文庫297、平凡社7巻365)

Quotation 6

すなわち四曼荼羅のうち、胎蔵界大日中に金剛大日あり。その一分心が大日滅心(金剛大日中、心を去りし部分)の作用により物を生ず。物心相反応動作して事を生ず。事また力の応作によりて名として伝わる。[…] いろいろの順序で心物名事の四つを組織するなり。[…] 右のごとく真言の名と印は物の名にあらずして、事が絶えながら(事は物と心とに異なり、止めば断ゆるものなり)、胎蔵大日中に名としてのこるなり。これを心に写して生ずるが印なり。故に今日西洋の科学哲学等にて何とも解釈しようなき宗旨(クリード)、言語(ランゲージ)、習慣(ハビット)、遺伝(ヘレジチー)、伝説(トラディション)等は、真言でこれを実在と証する、すなわち名なり。(明治36年8月8日、文庫338–339、平凡社390、往復書簡333–334)

Quotation 7

因はそれなくては果がおこらず。また因異なればそれに伴って果も異なるもの、縁は一因果の継続中に他因果の継続が鼠入し来たるもの、それが多少の影響を加うるときは起、(甲図。熊楠、那智山にのぼり小学校教員にあう。別に何のことなきときは縁。)(乙図。その人と話して古撃剣の師匠たりし人の聟ときき、明日尋ぬるときは右の縁が起。)故にわれわれは諸多の因果をこの身に継続しおる。縁に至りては一瞬に無数にあう。それが心のとめよう、体にふれようで事をおこし(起)、それより今まで続けて来たれる因果の行動が、軌道をはずれゆき、またはずれた物が、軌道に復しゆくなり。予の曼荼羅の「要言、煩わしからずと謂うべし」というべき解はこれに止まる。(明治36年8月8日、文庫340、平凡社391, 往復334)

Quotation 8

今日の科学は数を基として至ってたしかそうにいう、しかし数もまた定まらぬものなることをいえり。[…] 数というもの、今の科学者の思うほどたしかなものにあらず。事物を宇宙の一分一局にて測ることとなるが、実は事物によりて存するものなり。[…] 事物の根本にあらずして、その外を性質づける(クオリファイ) […] 数の外に確精なる事相なしというは早計もまたはなはだしというべし。今日の科学は数で量れぬものを度外視す。(臭、味のごとき、その学なし。数で計ることならぬゆえなり。されど、臭にいろいろあり。味にもいろいろあるなり) これにて今日の科学の不満足なるを知るべし。[…] 科学家の弊は万事科学科学いうて、科学で分からぬことをはうそのようにいうにあり。すなわち世界は器械にして、人間は動物なり、道徳もなにも虚偽なりというようなことになりたり。(明治36年8月8日、文庫344–345、平凡社393–395)

Quotation 9

故に、今日の科学、因果は分かるが(もしくは分かるべき見込みあるが)、縁が分からぬ。この縁を研究するがわれわれの任なり。しかして、縁は因果と因果の錯雑して生ずるものなれば、諸因果総体の一層上の因果を求むるがわれわれの任なり。(明治36年8月8日、文庫341、平凡社391–392)

Quotation 10

10を3にて除するに3.33333333 + 乃至幾億万回を重ぬるも極処に到らず、到らぬが、二回は一回より、三回は二回より実境に近くなる。3.3〜3を乗ずる(9.9)より3.33〜3を乗ずる方(9.99)が百分の九だけ10に近くなる。人のこの世にあるは安心を好む。安心は愉快の極なり。安心に等差あり。9.9の安心あり。9.99の安心あり。また9.999 + の安心あり。いまだ10に至らずといえども、安心深きものほど10の安心に近し。(明治35年3月22日、文庫234、平凡社7巻316)

Quotation 11

宇宙万有は無尽なり。ただし人すでに心あり。心ある以上は心の能うだけの楽しみを宇宙より取る。宇宙の幾分を化しておのれの心の楽しみとす。これを智と称することかと思う。[…] 今の学者が世界中のみな哲学者になる世を望むも、実は書籍多くよみ脳に不用のことをつめこむにあらず。自分の境に応じて宇宙無尽蔵の事相を一班なりとも心に考察して楽しむを望むに外ならざるべし。(明治36年6月30日、平凡社329–330, 往復書簡275)

References

Cook, Francis H. 1977. *Hua-yen Buddhism: The Jewel Net of Indra.* University Park, PA: Pennsylvania State University Press.

IAEA, INES. https://www-news.iaea.org/html/ines.pdf. Accessed 20 July 2015.

Īkura, Shōhei 飯倉照平, 2005. *A Japanese Translation of Minakata Kumagusu's English Treatises for 'Nature' Magazine* 南方熊楠英文論考[ネイチャー]誌篇. Tokyo 東京: Shūeisha 集英社.

IPCC Expert Meeting on Geoengineering. Meeting Report. (Lima, Peru: 2011). https://www.ipcc-wg2.gov/meetings/EMs/EM_GeoE_Meeting_Report_final.pdf. Accessed 6 August 2015.

IPCC Fifth Assessment Report (AR5). https://www.ipcc.ch/report/ar5/. Accessed 2 December 2015.

IPCC TAR Report. http://www.ipcc.ch/ipccreports/tar/wg3/index.php?idp=128. Accessed 29 November 2015.

Karasawa, Taisuke 唐澤太輔, 2015. *Minakata Kumagusu: The Limit of the Possibility of Japanese* 南方熊楠-日本人の可能性の極限. Tokyo 東京: Chūōkōron-shinsha 中央公論新社.
Keith, David W. 2000. "Geoengineering the climate: History and Prospect 1." *Annual Review of Energy and the Environment* 25.1: 245–284.
Masaki, Akira 正木晃, 2012. *Mikkyo* 密教. Tokyo 東京: Chikumashobō 筑摩書房.
Minakata, Kumagusu. 1893. "The Constellations of the Far East." *Nature* 48: 541–543.
Minakata, Kumagusu 南方熊楠, 1946. *The Complete Works of Minakata Kumagusu* 南方熊楠全集, vol. 9, edited by Shibusawa Keizō 渋沢敬三. Tokyo 東京: Kangensha 乾元社.
Minakata, Kumagusu 南方熊楠, 1971. *The Complete Works of Minakata Kumagusu* 南方熊楠全集, vol. 7. Tokyo 東京: Heibonsha 平凡社.
Minakata, Kumagusu 南方熊楠, 2010. *The Letters of Minakata Kumagusu Possessed by Kōzanji: For Dogi Hōryu* 南方熊楠書翰 — 高山寺蔵 土宜法龍宛 1893–1922, edited by Okuyama Naoji 奥山直司, Undō Hitoshi 雲藤等, and Kanda Hideaki 神田英昭. Tokyo 東京: Fujiwara-shoten 藤原書店.
Minakata Kumagusu Archive. http://www.minakata.org/. Accessed 23 November 2015.
Murphy, James M., David M. H. Sexton, David N. Barnett, Gareth S. Jones, Mark J. Webb, Matthew Collins, and David A. Stainforth. 2004. "Quantification of Modelling Uncertainties in a Large Ensemble of Climate Change Simulations." *Nature* 430: 768–772.
Nakazawa, Shinichi 中沢新一, 2006. *Baroque of the Forests* 森のバロック. Tokyo 東京: Koōdansha 講談社.
Oxford English Dictionary. http://www.oxforddictionaries.com/definition/english/mandala. Accessed 20 July 2015.
Rio Declaration on Environment and Development, 1992. http://www.unep.org/documents.multilingual/default.asp?documentid=78&articleid=1163. Accessed 15 July 2016.
Stevenson, Adlai II. 1965. "Strengthening the International Development Institutions." http://www.adlaitoday.org/articles/connect2_geneva_07-09-65.pdf. Accessed 12 March 2017.
Takemura, Makio 竹村牧男, 2015. *Japanese Buddhism: The History of Thoughts* 日本仏教 思想のあゆみ. Tokyo 東京: Kōdansha 講談社.
Taleb, Nassim Nicholas. 2007. *The Black Swan: The Impact of the Highly Improbable*. New York: Random House.
Taleb, Nassim Nicholas. 2015. *Silent Risk. Lectures on Probability, vol.1*. Descartes Publishing. The script can be found online: http://www.fooledbyrandomness.com/SilentRisk.pdf. Accessed 22 February 2017.
Tsurumi, Kazuko 鶴見和子, 1981. *Minakata Kumagusu: Comparative Studies on Preference of the Earth* 南方熊楠—地球志向の比較学. Tokyo 東京: Kōdansha 講談社.
Tsurumi, Kazuko 鶴見和子, 2001. *Minakata Kumagusu: The Thought on Suiten for the Paradigm Shift in the Future* 南方熊楠 — 萃点の思想 — 未来のパラダイム転換に向けて. Tokyo 東京: Fujiwara-shoten 藤原書店.
UNFCCC Paris Agreement, 2015. https://unfccc.int/resource/docs/2015/cop21/eng/l09r01.pdf. Accessed 15 July 2016.
Volcano Science Museum. 2000. "About Mt. Usu: Recent Eruptions after the 19th Century." http://www.toyako-vc.jp/en/volcano/about_usuzan/test5.html. Accessed 2 December 2015.
World Nuclear Association. http://www.world-nuclear.org/Features/Climate-Change/Climate-Change-and-Nuclear-Energy/. Accessed 29 November 2015.

Chapter 6
"Technique" and "Ethics" of Yuasa Yasuo from the Perspective of Asian Studies – An Introduction to Yuasa Philosophy

Sato Koetsu

1 Introduction

On the afternoon of February 1, 1989, a spacious lecture room at the University of Tsukuba was filled with students, faculty members, and various other people. The last lecture of Yuasa Yasuo 湯浅泰雄's (1925–2005) career at the university was about to begin. The lecture was entitled "Knowledge in the Age of the End of Philosophy." Yuasa, as usual, gracefully stepped up on the podium, dropping his eyes onto the manuscript as if to shy away from the devouring eyes of the massive audience. Then again, as usual, he started delivering his insightfully titled lecture to the hushed audience, in a voice so quiet and halting one could hardly hear it in the grand room. I too was present in the room as an assistant professor, but strangely enough, I hardly remember any of his lecture. Thinking back now, I cannot even be sure I was really there or not (therefore, my memory of this event, which is more

The original version of this chapter was revised: MS Mincho font was updated throughout the book for Japanese and Chinese characters. The correction to this book is available at https://doi.org/10.1007/978-3-319-59027-1_13

Sato Koetsu (✉)
Graduate School of Humanities and Social Sciences, University of Tsukuba, Ibaraki, Japan
e-mail: sato.koetsu.ga@u.tsukuba.ac.jp

Thomas Taro Lennerfors, Murata Kiyoshi (eds.), *Tetsugaku Companion to Japanese Ethics and Technology*, Tetsugaku Companions to Japanese Philosophy 1,
https://doi.org/10.1007/978-3-319-59027-1_6

like a lingering vision, is nothing more than an image reassembled from the fragments of memories left in my mind).[1]

Fortunately, the content of this lecture was published in the magazine *Japanology* shortly after Yuasa's retirement in a paper of the same title, and everyone was able to read it (Yuasa 1989b). Yuasa truly was a man of extensive reading and tenacious memory, as well as a prolific writer. His writing style was elegant yet simple enough for someone with less talent, such as myself, to understand his logic. Nevertheless, his writing condenses the *Espirit* of the East and West, giving his seemingly plain sentences unpredictable depth, which prompts one to be anxious of miscomprehending what is really going on. The writings of Yuasa are demanding in terms of the range of the material covered and the depth, which is a different kind of difficulty from comprehending his heard-to-hear lectures.

In this chapter, my intention is to select two of Yuasa's key essays, "*Knowledge in the Age of the End of Philosophy*" and "*Jungian Psychology's Inquiry into the Spirit of Modern Man –An Effort to Unify Eastern and Western Thought*" and to use them as points of departure for analyzing the framework of Yuasa's logical structure (Yuasa 1989a, b; Yuasa et al. 2001). Yuasa's special way of thinking about "Technique" and "Ethics" are explored and discussed in simple terms from the context of these two essays, which contain the best resources for providing an overview to the formative process and inherent concerns of Yuasa philosophy, as well as the goal he was aiming for. I hope to provide this overview through the use of Yuasa's own words and in a style that is possible to comprehend for readers who are not so familiar with his work.

[1]Before and after my time away due to participating in exchange programs from 1983 to 1985 at Chinese University of Hong Kong and Sun-Yatsen University in Guangzhou 廣州, I had the honor of spending a total of five and a half years as a graduate student and assistant professor to Yuasa. The temporary lack of a clear memory regarding this occasion may well be a result of the wealth of memories I have associated with the many lessons I learned from him on various occasions. Or it could be for a different reason. Certainly to this day there remains distinct in my memory the event entitled "10th Anniversary of Sino-Japan Peace Memorial Symposium" held from November 4–8 the previous year. That event, hosted by Yuasa, brought together *qigong* 氣功 (*kikō* in Japanese) researchers and practitioners of both Japan and China. I accompanied Yuasa for the duration of the event since I was appointed as a member of the executive committee and the head of the translator/interpreter team. Some of articles presented at the symposium and the performances of the Chinese *qigong* masters jolted and realigned my understanding (or so I believed) of Chinese Philosophy, History, Medicine, and so on. Looking back, I could very well have been in a state of confusion. After Yuasa's retirement from the University of Tsukuba in April 1989, he took a position as a professor of J. F. Oberlin University in Tokyo. Six months later, I also left my assistant's position at the University of Tsukuba and took a job as a researcher at the Center for Chinese Studies, National Library in Taiwan 臺灣. I then moved to the Institute of Modern History, Academia Sinica in Taiwan, and engaged in research on Chinese intellectual history until returning to Japan in March 1991. During this period, Yuasa consecutively published a series of books entitled *Jung and the East* (1989a), *Qi and Human Science* (edited and written, 1990), *The Body: Eastern Mind-Body Theory and the Modern Age* (1990), and *What is Qi?* (1991). He was kind enough to send all of these books to me. Along with my deep respect for Yuasa's novel ideas and research of unparalleled scale, I look back with nostalgia at the sense of delight and excitement I experienced in those days when introduced to these ideas resulting from Yuasa's contemplations.

2 Knowledge in the Age of the End of Philosophy

2.1 Formulation of Yuasa's Thought and His Historical Background

Yuasa started his lecture of this particular day by saying that he wished to discuss "how to think about the time in which he has been living, the present, from a philosopher's point of view," keeping in mind the ideas that inform post-modern and post-philosophical thought. Yuasa stated that he had experienced three distinctive time periods over the half-century during which he conducted his research. The first era starts from his birth and is punctuated by the Japanese defeat in the Pacific War. The second period starts with the post-war democracy and continues until the 1970s, which can be characterized as the era of ideology. The third spans from the 1970s to the present (1989, to be precise), which, in contrast with the preceding era, can be called the era of de-ideology.

The first era, namely the period during which Japan experienced defeat in the war, was an extremely shocking event for Yuasa's generation. Nevertheless, according to Yuasa, the political and social changes caused by that defeat, that is, the shift from pre-war nationalism to post-war democracy, brought about a transition within Japanese society and was not a major juncture in contemporary world history. Rather, the global change that has been happening since the 1970s – philosophically speaking, the quickening of the development of post-modern and post-philosophical thought –, during what Yuasa refers to as the third era, was far more important. Thus, Yuasa's analysis in this paragraph is extremely simplified. However, what has been implicated is considerably profound.

Needless to say, to those of us who are interested in the formative process of Yuasa philosophy, his first and second eras are equally important because the cultural and social factors of each of those eras were significant to the formation of Yuasa's thought. Before examining these three periods, let us first review the first half of Yuasa's life, about which he himself wrote in the postscript of the volume one of the "Complete Works of YUASA Yasuo" (Yuasa 1999–2003).[2]

Shortly after his entrance in April 1945 into the Department of Japanese History in the Faculty of Literature at Tokyo Imperial University, Yuasa was recruited into the student mobilization force. The war, however, came to an end in August, when he was a reserve cadet. Though he returned to the university by September, there were no faculty members left in the once-bridgehead of the *kōkoku shikan* 皇国史観, or the branch of Japanese history referred to as the Imperial View of the National History. As a result, he was forced to change his academic affiliation to the Ethics department 18 months later. He explained that the reason for his transfer was that "listening to the lecture of WATSUJI Tetsurō和辻哲郎 on the history of

[2]Volumes 1–12 and 14 published by Hakua Shobō (hereafter referred to as Hakua). Volumes 13 and 15–17 and Supplementary volume published by Being Net Press (hereafter referred to as BNP).

Japanese ethical thought made him think that he would be able to learn both Western philosophy and Japanese thought if he studied in the Department of Ethics" (Yuasa 1999, vol. 1: 494). Two years later in 1949, concurrently with Watsuji's mandatory retirement, Yuasa graduated after presenting a graduation thesis regarding the comparative consideration between Rickert's logicism and Bergson's intuitionism. Thereafter, while attending the study meetings of the successor of Watsuji, KANEKO Takezō 金子武蔵 (1905–1987), Yuasa developed an interest in economics[3] and re-entered the university's Department of Economics as an undergraduate in 1952. Following his graduation 2 years later, Yuasa proceeded to enter the Graduate School of Social Science as an Economics major. After finishing that major in 1956, he returned to the Ethics Department in the Faculty of Literature in order to work as an assistant professor. It was around this period that he developed a new interest in the issue of religion and ascetic practice experience.[4] He also began his study of Jungian psychology. Moreover, Yuasa accepted teaching jobs in both the junior and senior high school at Gakushūin University in 1961, and then later in 1966 took a position as an associate professor in Yamanashi University's Department of Education. After publishing *Modern Japanese Philosophy and Existentialism* (Ph.D. thesis, the University of Tokyo) in 1970, Yuasa successively served as a professor at Yamanashi University beginning in 1973, and in the Department of Literature at Osaka University the following year, and he spent a year as a visiting professor at the University of Indonesia's Literature Department in 1980. He received the post of professor in the Department of Philosophy and Thought at the University of Tsukuba in 1981. Leaving the key group of his written work for a later discussion, at this point I would simply like to close by mentioning his hosting of two important symposiums: the 1984 Franco-Japanese collaborative international symposium in Tsukuba 筑波 titled "Technology and the Spiritual World" and the 1988 Sino-Japanese 10th anniversary peace memorial symposium titled "*Qi* and Human Science."[5]

[3] According to Yuasa, his interest in economics was motivated by his own belief that one could not understand society without understanding economics. On the other hand, his experience in the social unrest that followed Japan's defeat in WWII made him lose any sort of trust in politics, whether it be associated with the left or the right (ibid.: 498). Moreover, he claimed that no serious philosophy or thought could be learned from politics or ideologies (ibid.: 496), and that ideology was nothing but a type of corrupted form of traditional philosophy enmeshed in politics (Yuasa 1989b: 122)

[4] Yuasa was born in the house of an associate leader of a Shintō-affiliated religious sect, *Hitonomichi Kyōdan* (literary "Church of People's Way," the predecessor of the present "Church of Perfect Liberty"): in his later years, he encountered MOTOYAMA Kinue 本山キヌエ, who was the originator of Tamamitsu Shrine sect, and he studied self-cultivation and Yogic meditation methods from MOTOYAMA Hiroshi 本山博, the chief priest of that sect. All of these factors motivated Yuasa to become more deeply inclined to devote himself to religion.

[5] For Yuasa's career history after this position as an assistant professor at the University of Tokyo, I referred to the "Chronological Record of YUASA Yasuo," included in *Mind-Body Knowledge—Inheritance and Development of Yuasa Philosophy*, BNP, 2015. This book received attention as the latest collection of research papers on Yuasa philosophy, and it provided me with many suggestions that assisted me in the writing of this thesis.

Yuasa's progress as a researcher started in what he called his second era. He offered three keystones for the analysis of his own research. The first is the tradition of Chinese and Indian wisdom with a focus on Japanese Thought.[6] The second is a particular perspective coming from psychology and psychiatry, namely Jungian psychology.[7] According to Yuasa, the reason why he was interested in Jung was that it provided a way to translate the tradition of the Eastern wisdom (e.g., Yoga, Taoism, Buddhism and Divination) into a form applicable to the psychological and spiritual issues of modern people. Yuasa himself had been examining the characteristics of the tradition of wisdom and knowledge in both the Western and Eastern worlds, under the guidance of his Jungian studies. What was particularly important to him was the fact that Jung's insights regarding Western wisdom pointed outward from the self, whereas Eastern wisdom pointed inward. Yuasa treated his Jungian studies as an opportunity to broaden his examination of the mind-body dichotomy and the relationship between the material and the spiritual. This work led to the third keystone, the mind-body theories. His representative work on the subject is *Shintai: Tōyōteki shintairon no kokoromi* (published by Sōbunsha in 1977) and its English translation. *The Body: Toward an Eastern Mind-Body theory* was translated in English and helped Yuasa's name become widely known among scholars in Europe and the US (Yuasa 1987b).[8] His interest in this issue was based on his desire to find a clue that would allow him to overcome the tradition of knowledge and wisdom that had developed since Descartes (or, we could even say since the Greeks), which has been grounded in the mind-body distinction or matter-mind dualism. According to Yuasa, the Eastern tradition of wisdom does not include within it any form of dualistic confrontation, but rather it aims for a high level *theōria* that can be attained solely through practice, namely, a practical recognition of the inner world of the mind (Yuasa 1989b, p. 130).

---

[6] Excluding Yuasa's translations and co-author works, his writings include the following: on India; *The Ethical Thought of Indian Epic Poem*, 1957; on China; *The Ethical Thought of Mohism*, 1957; *The Ethical Thought of Legalism*, ibid.; on Japan; *Moral Emotions of Ancient People*, 1957; *Ideals of the Ancient Nation*, ibid; *Mysticism and Ethics of Kūkai*, 1960; *The Ethical Thought of the Ancient Nation*, 1966; *Philosophy and Existentialist Thought in Modern Japan*, 1970; *The Spiritual World of the Ancient People*, 1980; *The Religious Consciousness of the Japanese People*, 1981; Watsuji *Tetsurō—The Fate of Modern Japanese Philosophy*, ibid.; *Thought in Japanese Mythology*, 1983.

[7] His other representative writings include such titles as *Jung and Christianity*, 1978; *Eastern Religions and Depth Psychology*, 1979a; *Jung and the European Spirit*, 1979b; *The Deeper Layers of Eastern Cultures*, 1982; *Literary Fragments on Views of Nature and Self-Cultivation Theories in Esoteric Buddhism*, 1984; *The Psychology of Eastern Meditation*, 1983. *The Psychology of History and Myths*, ibid.; *A Study on Synchronicity*, 1987a.

[8] Translated by NAGATOMO Shigenori 長友繁法, and Thomas P. KASULIS. Edited by Thomas P. KASULIS. Other translated works include: *Mind-Body Relation Theory and Issues of Ascetic Practice*, 1982. *Problems around the Eastern Body Theory*, 1983. *Literary Fragment about View on Nature and Ascetic Practice Methodology of Esoteric Buddhism*, 1984. *Qi Ascetic Practice and Body*, 1986. As an excellent researcher of Japanese Intellectual history, Professor Kasulis has also authored a comparative study of Japanese and Western thought entitled, *Intimacy and Integrity: Philosophy and Cultural Differences* (Kasulis 2002)

I intend to discuss in greater detail these various issues mentioned thus far in the following sections.

For the sake of discussion hereafter, it is necessary to clarify the exact meaning of the lecture title, "Knowledge in the Age of the End of Philosophy." The "end of philosophy" lies on the second stage of Yuasa's periodization, which also coincides with the demise of the era of ideology. The question is, then, when did the era of ideology come to its end, and how does that relate to "philosophy"?

2.2 *The End of Philosophy*

The start of the era of ideology, in Yuasa's mind, is to be found well before the era of imperialism in the late-nineteenth century. World War II was followed by an era flooded with ideologies, including communism, socialism, and democracy. Not only did it influence politics and the economy, the world of philosophy was deeply affected by these movements. Existentialism was positioned at the center of these trends, and Jean-Paul Sartre, who had connected it with Marxism, held a central position in the philosophical world at that time.[9] Yuasa asked himself why such a

[9] Amano Teiyū 天野貞祐 (1884–1980), a philosopher and professor at Kyoto Imperial University who served as the Minister of Education after the war, has left us with a rather interesting episode on this topic. Amano studied at Heidelberg University from 1922 to the following year. When Amano returned to Heidelberg in 1958, a party was held at a hotel in Constance and there Heidegger introduced him to the group from the Berlin Academy of Arts with the following explanation: Amano was a close friend to the Japanese philosopher Kuki Shūzō 九鬼周造 (1888–1941). When Kuki came to Paris, he hired a young man to help him learn the French language. In return for his French lessons, he taught the young man the existential philosophy of Heidegger. This young man later became known as Sartre, the famous philosopher. Therefore, we can say it was the Japanese professor Kuki, who first introduced the Existentialism philosophy of Heidegger to France (Amano Teiyū, "*Jōhin de Seijitsu na Gakusei wo* 上品で誠実な学生を" (A decent and sincere student), "*Kokoro no Gunzō* 心の群像" (Images in the mind), edited by Kaji Ryūichi 嘉治隆一, Kawade Shobō, 1962: 207). This episode may contain a certain amount of embellishment by Heidegger, which grew out of his admiration for Kuki, but still is an interesting article that indicates the achievement of not only Kuki but also many Japanese scholars who were eager to learn about Western philosophy.

Further back in time, the person who made the famous statement, "There has been no philosophy in my country, Japan, until now" was Nakae Chōmin 中江兆民 (1847–1901). In Japan, he is known as the person who introduced Jean-Jacques Rousseau to Japan. In the course of the modernization of Japan, ever since the Meiji Revolution in 1868, there has been a clear tendency to move from just taking in Western Philosophy to the awakening of the originality of Japanese thought in comparison to the West. Kuki Shūzō was one of those who engaged in this approach, and his work entitled *Iki no Kōzō* (*The Structure of "Iki* いき"), 1930, displayed the achievement of his Heidegger-influenced analysis and contemplation of the aesthetic sensitivity found in the culture of Japan's common people, all viewed in the context of phenomenology. While rapid modernization took place in fields such as politics, economics and social structure, many excellent philosophers started to emerge from the generations that came before and after Kuki's; they included Nishida Kitarō 西田幾多郎 (1870–1945), Tanabe Hajime 田辺元 (1885–1962), Watsuji, and Amano. Their understanding of Western thought had grown so much deeper that it could be easily distinguished from that of an earlier period in Japanese modernization.

situation had developed to begin with, and came up with the following insight. He suggests that at this time people experienced the collapse of the social order that had been stably functioning for centuries, and this left them in a psychologically blank state, or better put, a state of collective anxiety in which they did not know what would happen next. Ideology simply slipped into this, void and filled it up. Moreover, this ideology was nothing more than intellectual concepts or fantasies, the tangible context of which was beyond the realm of people's full understanding. The truth is, he points out, that because these ideologies could not be fully understood or actualized in concrete terms, space was created for people to fantasize, charismatic leaders to emerge in politics, and thinkers such as Sartre to take center stage in philosophy. Yuasa continues to argue that the end of this era – which is suggested by the emergence of post-modern, anti-philosophy and deconstruction – is symbolized by an incident in 1960s, the argument between Sartre and Claude LÉVI-STRAUSS.

This argument famous throughout the Western world considered Sartre's criticism of Lévi-Strauss in his *Critique de la raison dialectique* (*Critique of Dialectical Reason*) and Lévi-Strauss's response to Sartre in the last chapter of his *La pensée sauvage* (*The Savage Mind*) (Lévi-Strauss 1962; Sartre 1960). The dispute clarified the viewpoints of the two, and became an opportunity for Structuralism to receive sudden attention as a transcender of Existentialism.

Based on Yuasa's understanding, the argument started with Sartre's criticism of Lévi-Strauss's anthropological research. Lévi-Strauss argues that there is no essential difference between primitive and civilized societies. Both of them have their own systems that employ certain symbols, which determine their social structures. These symbolic systems naturally rule over the behavioral patterns of people. This argument puts Sartre, who adopts a Marxist historical viewpoint that takes as its premise progress, class struggle and revolution, in opposition to Lévi-Strauss. Therefore, Sartre criticized his anthropological studies as being conservative. Lévi-Strauss, on the other hand, retorts that the fundamental core of Sartre's thinking is Eurocentric. In addition to it, Yuasa further points out that Lévi-Strauss identified the culture of Indio as a primitive civilization destroyed by European colonialism, and that this was a historical sin committed by Western civilization.

Certainly, there is room for debate on the nature of argument between Sartre and Lévi-Strauss, as well as Yuasa's understanding on it. However, perhaps most of the intellectuals today will accept as a fact that the argument concluded with Sartre's complete defeat, and resulted in Existentialism becoming a thing of the past.

On the contrary, I would like to point out two reasons why this argument lacks significance in terms of Yuasa's work. First, Lévi-Strauss himself revealed his sympathy for the Marxist historical viewpoint, one that is similar to that of Sartre, in his criticism of Sartre in *La pensée sauvage*. Thus, Marxism, in and of itself, could not be the core point of contention in their argument. Moreover, while using the expression

---

In this regard, Yuasa often used the term "Japanese Philosophy," which is seldom used in contemporary Japan by other researchers specializing in intellectual history. It is another indication of his enthusiasm and self-confidence in putting the thought of the East and the West on equal ground and then integrating them.

"the era of ideology," Yuasa asserts that, philosophically speaking, ideology is nothing but "a type of corrupted form of philosophy enmeshed in politics" (Yuasa 1989b, p. 122). This suggests that Yuasa was clearly skeptical towards ideology such as Marxism or other ideologies, for these were issues he couldn't care less about – except maybe in terms of their respective positions in intellectual history.

At this point, let us look back on Yuasa's background, whether it proves relevant or not, just to reconfirm the environment in which he grew up as a child. He probably was raised in an intensive religious environment under the influence of his family, especially his father YUASA Masao 湯浅真生,[10] who was one of the associate leaders (virtually the second highest rank in the sect) of *Hitonomichi Kyōdan* ひとのみち教団. He was born and raised in such an environment, which was well removed from Marxism, and stayed there his entire life. His religious sentiment, which was cultivated from childhood, certainly was one of the major influences that became the basis for his deep interest in and understanding of various religions worldwide, such as Japanese Buddhism and Shintō, Chinese Buddhism, Taoism, Yoga in India, as well as Christianity, including Gnosticism. At the same time, the political oppression towards *Hitonomichi Kyōdan* in 1937 was a tragic event that threw Yuasa and his entire family into an abyss. It is not difficult to image that the event cast a shadow over his mind, from his childhood into adolescence and remained with him to the end of his life. Such a formative experience in his youth must have been the reason for Yuasa's separation from politics and political science, both directly and indirectly.

In light of this history, then, we cannot help but wonder why Yuasa would venture into an analysis of Sartre and Lévi-Strauss argument. I would speculate that the one of the central reasons why Yuasa showed such an interest in this argument was that Lévi-Strauss focused his attention on that which was outside of Western philosophy (or Western world). In other words, he put himself there, on the outside of the Western civilization, and performed an introspective analysis on the overall Western culture, including philosophy. Later I would like to discuss this in greater detail, but what he was dealing with was related to the larger theme Yuasa was aiming for: the unification of the wisdom of the East and West. Another reason could be that, while Yuasa could not find any common term between his ideological interest and the philosophy of Sartre, he was able to discover firm fundamental connections, under the influence of Jungian psychology, in between depth psychology and the anthropology of Lévi-Strauss in the area of methodology and hermeneutics, such as sexuality, symbols, imagery and so on as seen through the ancients. Traces of his discovery can be seen in Yuasa's studies on Japanese mythology.

---

[10] I was able to get a great deal of valuable information on aspects of Yuasa's life from a gift presented by MARUYAMA Toshiaki 丸山敏秋 entitled "MARUYAMA Toshio and YUASA Masao," *Bulletin of RINRI Institute of Ethics*, vol. 20, 2011: 10–52. MARUYAMA Toshio 丸山敏雄, the grandfather of the board chairman of the RINRI Institute of Ethics, Toshio, and Masao both played important roles in the education and research in the religious organization as intellectual elites. For that reason, they were put on the firing line along with the founder of the organization against the oppression from the fascist government of the time.

In Yuasa's phrase, the victory of anthropology over philosophy also meant the end of the era of philosophy. As abrupt as it is at a glance, this expression could also be translated into marking the end of that civilization in which the form of discipline called "philosophy" was at its highest expression (or, in which knowledge was at its highest order) of all the disciplines.[11] If so, what was called "the end of philosophy" was actually "the end of Western philosophy" in its simplest form. More strictly speaking, it was Western philosophers themselves who felt they were experiencing "the end of Western philosophy." While referring to the names of Heidegger and Wittgenstein, and paying more careful attention to the works of Richard Rorty (1979, 1982),[12] Yuasa explained his reasoning for concluding, for the time being, that the contemporary world is entering a post-philosophy era, that is, an era of civilization in which the discipline called "philosophy" had come to an end. However, this post-philosophy era will not be analyzed in this article since our main focus is to clarify the characteristics of Yuasa's philosophy.

2.3 *Wisdom of* theōria *and Wisdom of* prāxis

The question is, then, how he reached such a perception. Yuasa's own explanation of the reason indirectly reflects the characteristics of his own philosophy well. What is especially remarkable was that Yuasa viewed the superiority of the "wisdom of *theōria* (*θεωρία*)" over the "wisdom of *prāxis* (*πρᾶξις*)" as the basic structure of Western philosophy.[13] It is specifically described as follows.

Generally speaking, the history of Western philosophy goes back to the days of Socrates. The philosophy of Socrates places the awareness of the "ignorance of knowledge" at its core. According to Yuasa, this famous proposition was first voiced by Socrates, who brought awareness to the ignorance of people like the Sophists, which were well regarded for their objectivity and empirical knowledge about nature and society, but were ignorant about *psychē* (*ψυχή*), or the human soul. This coincides with the birth of philosophy. On the contrary, the idea of having *theōria* over *prāxis* was actually established by Plato and Aristotle in ancient Greece. Modern Western philosophers such as Descartes and Kant also defined "philosophy" as "the wisdom of *theōria*, or a higher order that integrates prāxis, or objective and empirical knowledge." Thus, the tradition of Greek-Western philosophy was sustained without interruption until the nineteenth century. "The end of philosophy" put another way, can be described as "the independence of science from the philosophy that with the modernity of West, and inversion in the relationship between

---

[11]Yuasa utilizes the German term "*Wissenschaft der Wissenschaften*" in describing "the highest knowledge of knowledges" (ibid.: 121).

[12]The expression "the Age of the End of Philosophy," which Yuasa used, most likely came to him as a hint from a series of Rorty's work, including *Philosophy and the Mirror of Nature* (1979).

[13]Yuasa uses the term *prāxis* in a broad sense including the meaning of *poiēsis (ποίησις).*

science and technique that took place in the twentieth century both signify that the end had finally arrived for the era of the civilization that was ruled by such patterns of thought" (Yuasa 1989b, p. 129).

It is obvious that there are two key factors at the foundation of Yuasa's understanding of Western civilization: one is the concept of dualism that separates mind and material as seen in Greek-Western civilizations; and the other is the emergence of a new and powerful phenomenon, namely, the development in the twentieth century of science and technology.

As it is widely known, Greek philosophy adopts a standpoint that separates form (*ἰδέα*, *idea*) and matter (*ὕλη*, *hylē*), a view that appears to have originated in Orphicism. In particular, Plato is known to have set the world of idea outside the matter-filled world of reality; he declared that witnessing it is what *theōria* is about and only philosophers could reach such a state. During the Middle Ages in the West, Plato's philosophy was incorporated into the dualism of body and soul (mind) in Christian theology, in which, needless to say, soul (mind) was more highly valued than the body. Yuasa points out that Descartes did place mind and material in an equal position, but for this very reason he had to define them as separate and unrelated aspects of existence. Thus the separations we see after the modern period between idealism and realism, spiritualism and materialism, and consequently philosophy and empirical science all originated here. Simply put, philosophy has taken the form of metaphysics and become the wisdom about material, taking the form of natural science. This sums up my understanding of Yuasa's analysis, which still needs more supplementation on the relationship between science and technique from the context he provides.

Yuasa viewed the scientific revolution from Copernicus to Newton as something that contributed nothing to the improvement of the social position of either technique or the practitioners of technique, the technicians. Technicians – whose roots, Yuasa claimed, could be traced back to the medieval alchemists (yet another obvious influence from Jung) – included such people as Thomas Edison and the Wright Brothers. Based on the social norms, they were placed in a subservient social position. To put it in other way, up until the nineteenth century, "professors and scientists were the people who thought and categorized things based on experience and observation, whereas the technicians in the field were the people who applied and implemented their ideas to make use of them." (Yuasa 1989b, p. 124). Thus, science in this stage became a sub-branch of philosophy and functioned also as the wisdom of *theōria* as much as philosophy did. While René DESCARTES was a mathematician, Isaac Newton, a physicist, and Adam SMITH, an economist, they were, on the other hand, simultaneously representative modern philosophers of rationalism, naturalism and ethics (Yuasa 1989b, p. 121). However, as we entered the twentieth century, in various fields such as physics (quantum, space, astronomy, robotics, etc.), medical science, biology, and so on, technique has made astounding improvements (and been subdivided at the same time). This has caused science to become incapable of fully assuming the role of organizing the concepts behind the technique. Not only that but science has become in many ways subordinate to technique. With these supporting examples, Yuasa insists that "the root cause for the end of the era of philosophy was the technological development" (Yuasa 1989b, p. 124). The reasoning for his argument is very clear: science was once viewed as the wisdom of *theōria*, but it mixed with technique, which used to be the wisdom of *prāxis* until modern times, and together they were transformed into science-technology, which in the end created a new ground to stand on, one that was different from "philosophy." From my point of view, "science" today merely is an acronym for "science-technology." It may be no exaggeration if we were to say it is, in fact, a "technology." In short, the reality of science is that it is absorbed into technology if not hidden behind it. On the other hand, "philosophy" – once viewed as the reigning "supreme wisdom," the fount of all disciplines – seems to have lost what it was seeking to "look at" (*θεωρία*, the verb form of *theōreō* (*θεωρέω*) stands for "to look at, to view"). Judging from the point of view discussed above, there is nothing but the end of conventional philosophy, which had been serving as a means of differentiating the wisdom of *theōria* and *prāxis*, and giving superiority to the former. At last that is how I would interpret the words of Yuasa, which are available in English only in this thesis.

Regarding this point, Yuasa falls in with the previously noted Rorty, who considers Francis Bacon (1561–1626) as the better representative of modern philosophers, pointing out that naming Descartes and Kant as the primary representatives was a mistake. Yuasa states:

> What Bacon meant by saying "*Ipsa scientia potestas est* (knowledge is power)" is that knowledge is primarily inseparable from the power gained through experience, that is technique. In other words, the way of thinking, in which *theōria* and *prāxis* are differentiated with superiority given to the former, itself is an indication of the possibility that the modern European had been living in some sort of a great misapprehension about the changes in world history that had begun in the modern Western world. In the world of modern philosophy, epistemology (i.e., contemplation on the form of knowledge) is often considered the supreme subject. To clarify the methodological premise of empiric science is also a meaningful subject of academic theory. The thinkers of the postmodern era, however, have asserted that this kind of basic attitude of modern philosophy itself was questionable. (Yuasa 1989b, pp. 124–125)

He continues on to state that, "while the modern rationalists represented by Descartes or Kant are representing the wisdom of *theōria* for the elite class, Bacon's footing seems to represent the technical view on nature science alchemists, which is the wisdom of *prāxis*. And the true driving force of the modern world seems to rest more on the latter than the former." (Yuasa 1989b, p. 125)

What is significant in relation to this statement is his determination that the differentiation between the wisdoms of *theōria* and *prāxis* and the placing of superiority on the former only happened in the West and no other civilization. Needless to say, Yuasa's assessment is a reflection of his own ideological characteristics. Aside from that, it prompts us to reexamine modern day rationalism. That is to say, this argument is directly connected to Yuasa's view on how the historical source of technique had strong connections with religious ceremonies in various ethnic groups, and is significant as an insistence that urges us to reconsider the issues of body or mind-body, which modern rationalism has often overlooked.

In the following, rather long quote, Yuasa discusses these issues more directly.

> For a human to think based on reason means to think within the reality of the world –within history and culture, to be more specific – and does not mean to think in a kind of vacuum that extends beyond reality. Modern-day people believe reason or logic to be something universal, something beyond the diversity of culture and history. Reason, within the context of history or the tradition of ethnic cultures, always functions in tangible form. Although reason and logic may be universal to humans, it can only function when it is put within historical circumstance and cultural diversity. Modern Westerners have not been aware of this fact. Moreover, inside reality, humans always act first in order to live. Thinking does not come first. In other words, the wisdom of *theōria* cannot be separated from *prāxis* to begin with, and at the foundation of *theōria* lies *prāxis*, which supports it. When these were separated and people began to believe that there could be a world-viewing wisdom that issued from the position of God, then "philosophy" as the wisdom of *theōria* was born. (Yuasa 1989b, p. 126)

It needs to be clarified here that Yuasa is not thinking of an approach provided by *theōria* as unacceptable or worthless. He deliberately added an explanatory note following this paragraph, stating that the wisdom of *theōria* is the ideal or the goal for wisdom, and that we must be aware of the fact that ideal can never be completely achieved.

2.4 Technique and Technology

Furthermore, Yuasa described "technology" as follows:

> When we say "technology," we all instantly associate it with the large-scale technology we see today. But if we look back in the history of human beings, we discover that there existed various unique technologies (*technē*: *τέχνη*, technique) in numerous ethnic cultures. Human-invented technology starts with the use of fire and stoneware in the remote past of the Stone Age. Technology also includes the creation of earthen wear, cookery of food, textile, hunting, art-making skills, religious ceremonies and medicine. These technologies show the traditions of culture based on the differences in history and climate among each ethnic group.
>
> In such case, one significant point I would like to stress is that, in any ethnic group, the historical source of technology was always tightly connected with religious ceremonies. This is evidence of the fact that technology is naturally inseparable from the human body... Since technical knowledge is a kind of empirical knowledge gained through the mastery of "*waza* 技 (technique in Japanese)," it cannot be considered separately from the issues of body. For example, in Japan, the tradition of "*waza*" seen in the fine arts, martial arts or craftsmanship derived from religious ascetic practices that are still actively maintained and utilized today. There, the mastery of "*waza*" does not only consist of physical training, but it also means to train the person's mind and improve the personality... In short, technique originally requires psychosomatic correlation as a prerequisite. Thus, the training for "*waza*" means training for the mind through physical training. (Yuasa 1989b, p. 127)

As we have seen, Yuasa makes clear distinctions between "technique" and "technology." He had a high opinion of the modern significance of technique, which seemed to have been buried behind large-scale technology. He valued it in the context of issues related to body-mind theory, as well as from an ethical viewpoint. This indirectly suggests his opinion about how issues of technology in the modern world are directly connected to various ethical problems in today's society.[14] These issues will be discussed again later in this chapter.

Thus, Yuasa's answer to the question of what is the basic difference between technique and technology is evident. "The masterpiece of Leonard da Vinci will never appear again" (Yuasa et al. 2001, p. 25). The technique of da Vinci has vanished long with him when he died. Yuasa points out that technology does not belong to a person and anybody can have benefit from it, but that is not the case with "technique." That is what makes "technique" valuable and the reason for its deep connect to people's minds.

I myself recall the first time I stood in front of the *Mona Lisa* (of the bullet-proof glass in front of it) one autumn some 20 years ago. I felt a kind of deep emotion that affirms these words of Yuasa. Even though this portrait with its mysterious charm has been the subject of considerable conjecture regarding the circumstances of its creation, no one has objected to the fact that it was created by the proficient skill and mature soul of da Vinci in the early sixteenth century. Moreover, what has made it possible for modern people to see this piece approximately 500 years after its creation are the technique and souls of numerous other people, famous and nameless, involved in its restoration over this time span. On the contrary, when we look at those people with no self-control who have thrown acid or rocks and have sprayed paint on it, we realize they are poles apart from those who have worked to preserve it. It is only natural then for us to feel respect from the bottom of our hearts towards the noble human spirit or the personality of people who ensured its safety through the frequent ravages of war and who helped to maintain this heritage of the human race with their careful and humble restoration work. There certainly is a reason to view technique and personality as a continuum. In the meantime, in order to semi-permanently preserve the *Mona Lisa*, we must utilize the latest technology of the time. Yuasa's metaphor is ingenious in this sense. We should go farther in our examinations of the complexity of the issues we face today, that is, the reality regarding the relationships between human and society, technology, politics and economy.

Returning then to the main subject, Yuasa continued on to state that, "modern-day wisdom, which is grounded in the mind-body dualism, emphasized knowledge about the material world as well as the mechanical technology that controls and dominates it. But it does not recognize any technology about mind" (Yuasa 1989b, p. 128). According to Yuasa, prior to the modern era, there were traditions of empirical wisdom that trained the body and mind (soul) at the same time, or, in short, self-cultivation practices. These can be understood as traditional cultural heritages seen in, for example various Eastern religions. Modern-day depth psychology and psychosomatic medicine recognize their significance. The most significant part is that the tradition of Eastern wisdom does not separate the knowledge acquired through *theōria* and *prāxis*. Specifically, it aims to reach the higher wisdom of *theōria* that is gained through *prāxis*. It is a methodology that focuses on the technical wisdom of

[14] In Yuasa's own words, "That which influences the spiritual and psychological problems of contemporary people is technology, not scientific theory" (Yuasa et al. 2001: 23).

knowledge available to people that will allow them to guide and control the self. This is, he insists, the kind of higher wisdom that is needed today.

As preface to the conclusion of his lecture, Yuasa suggested that the search for the way wisdom or knowledge should be in the Age of the End of philosophy will be the work of the thinkers hereafter, and by reviewing the tradition of Eastern wisdom, they may be provided with one of the clues (not the only clue) that can assist in their search for the way in which knowledge and wisdom should be understood today. His lecture concluded as follows:

> Today, if there is a universal form of knowledge and technique related to the material world as we see it, it may be possible to conceive of the existence of a similar universal wisdom and technology in the world of mind. I suspect that this task may be related to a knowledge and wisdom that will allow us to unravel the meaning inherent in the traditions of the mind found in various ethnic groups around the world as well as the tradition of civilization. (Yuasa 1989b, pp. 130–131)

Here, the universal knowledge and technique regarding the world of the mind, according to Yuasa's assertions, is best understood as implicating the integration at a higher dimension of "knowledge" the traditions of knowledge developed independently in the cultures of the East and the West. By reviewing the progress of his own research in a very humble tone, Yuasa elucidated some of the future prospects for his research and, at the same time, declared that the subject of integrating the wisdoms of the East and the West would be his lifework.

Assuming that my understanding of Yuasa's work is free of error, I would say that what he truly desired to discuss here was not the end of philosophy but actually its rebirth. *Between Religion and Science* (*Shūkyō to Kagaku no Aida*) was published by Yuasa 4 years later (Yuasa 1993). Included in the introduction was the complete version of this lecture "Knowledge in the Age of the End of Philosophy." The title, however, was revised to "The Rebirth of Philosophy," which further clarified Yuasa's intention. What lead Yuasa to such a state of mind was Jungian psychology.

3 Jungian Psychology's Inquiry into the Spirit of Modern Man – Toward an Integration of Eastern and Western Thought

3.1 Wisdom of the West and the East

As revealed in his retirement memorial lecture, the reason why Yuasa had been interested in C. G. JUNG from the 1950s was because "Jung was pointing toward a new way of understanding the tradition of Eastern wisdom, as seen in Yoga, Taoism and Buddhism, which he saw as in line with the mind-body issues of contemporary people" (Yuasa 1989b, p. 130). What is fascinating about this fact is that under the guidance of Jung's thought, in which Western knowledge is defined as moving outward-from-self in contrast to the Eastern approach that proceeds inward-to-self, Yuasa confronted the essence of the dualism between materialistic and spiritual wisdoms, and finally arrived at his conclusion that the tradition of the Eastern approach to knowledge and wisdom does not contain the dualism or even the idea of separating *theōria* and *prāxis*.[15] In other words, Jung's thought had lead him to realize that there was a unique tradition of *theōria* and *prāxis* in the East as well, even though Easterners themselves had almost forgotten about it. At the core of his realization were the mind-body issues that had emerged from his contemplations on the recesses of ascetic practice in the East. Yuasa's well-known book *Body* or *Shintai* came out of such concerns. Thereafter, Yuasa consecutively published a number of works in the related field. Some of the titles of these works include *Jung and the East* (1989a), *Qi and the Human Science* (1990, authored and edited), *The Body: Toward an Eastern Mind-Body Theory* (1990), and *What is Qi?* (1991). Two of the most important works from this period that condensed his thought on the Eastern philosophy are *Shintai no Uchūsei: Tōyō to Seiyō* (*Cosmic View of Body: East and West*) (1994) and *Kyōjisei no Uchūkan: Jikan, Seimei, Shizen* (*Synchronicity of the Vision of the Universe: Time, Life and Nature*) (1995). Out of all of these invaluable works, only *The Body* (1990) has been translated into English so far. In the former, using the key concept of the "body," Yuasa dealt with the relationship between human beings and nature as a correlation of the micro and macro cosmos. In the latter, he discussed "synchronicity," one of the key concepts in late Jung's psychology, in relation both to "time" and "nature" in *The Book of Changes* and mind-body theory. Unfortunately, it is impossible to go into the details of each of these works and to consider the individual themes found in this thesis. While keeping in mind the arguments developed in these works, I would like to focus on how Yuasa in his final years viewed the integration of the wisdoms of the East and West through his encounter with Jungian Psychology, utilizing his 2001 work *Jungian Psychology's Inquiry into the Spirit of Modern Man*.

Being one of his last authored works, this particular piece has not been translated into any European language as far as I know. Therefore, it may be helpful to summarize it. The following is the construction of the thesis.

1. What Jung learned from the East
2. Arrival of the post-modern era
3. Considering various contemporary ethical issues from a psychological point of view
4. Inundation of sex in the contemporary world
5. Psychology of *The Secret of the Golden Flower*
6. Bioethics and the ethics of connection
7. Human as the existence within nature

[15] Joseph Needham (1900–1995) shared a common understanding of this issue with Jung. Yuasa had become familiar with his thoughts on how China, Japan, and other East Asian cultures naturally lacked the idea of a dualistic separation between the mind and matter based on his reading of the Japanese translation of Needham's work entitled *Science and Civilisation in China*, as well as other studies by him.

As he points out, the general theme for his thesis focuses on "What Jung learned from the East?" and concurrently he contemplates "what Jung expected from the movements in contemporary society, beginning with what he found in America, and what he felt about them" (Yuasa et al. 2001, p. 12).[16]

In the first chapter entitled "What Jung learned from the East," Yuasa clearly states that the two works that influenced Jung the most were *The Book of Changes* and *The Secret of Golden Flowers*.[17] The following statement by Yuasa indicates the importance of taking up *The Book of Changes* as the subject for his thesis:

---

[16] Incidentally, readers may wonder why Yuasa so often made mention of not only the encounter between Jung and the East but also American society. Put simply, at the innermost recesses of Yuasa's thought, he speculated that Jung's psychological analysis of American society during his visit to the States in 1909 and 1912 would become applicable to European society in the near future and ultimately to Japanese society in much the same way. "The fact that historical changes are always initiated in advanced nations is yet another truth about the history of the human race. We must pay attention to the signs of changes quickening in this nation (America)" (ibid.: 44).

[17] The two books listed here, *The Book of Changes* and *The Secret of the Golden Flower,* are the *I Ging, Das Buch der Wandlungen*, 1924, and *Das Geheimnis der goldenen Blüte, ein chinesische Lebenbuch,* 1929, both by Richard Wilhelm (1873–1930). Wilhelm is considered to be the forefather of Chinese studies in Germany. Since his vestiges and these two translated books are already well known throughout the West, it may not be necessary to give detailed explanations. However, for the convenience of later discussions, some supplementary information on specific matters needs to be provided.

First of all, Wilhelm's notation on the Chinese title, *I Ging,* is translated into the *I Ching, or Book of Changes* in the English translation done by Cary F. Baynes, which follows the rules used since the latter half of the 19th century and which are referred to as the Thomas Wade (a.k.a. Wade-Giles) system. However, today in most of the East Asian world, people rely on the Pinyin System established by the Chinese Government in 1956. With Confucius Institutes spreading throughout the world, there is no need to return to using the classical mode of expression. This chapter also follows the Pinyin system for the notation of Chinese words, therefore *I Ging* and *I Ching* are presented as *Yijing* 易經. Strictly speaking, its source book in Chinese is titled *Zhouyi* 周易. *Zhouyi* consists of a textbook for divination, *Yijing* and its Confucian instruction manual, *Yizhuan* 易傳 (also known as *Shiyi* 十翼, or *Ten Wings*). Though uncertain folklore is the only clue to what occurred during their years of creation, the closest estimation suggests that *Yijing* was established during the Chunqiu 春秋 Period, whereas *Yizhuan* was produced until the Former Han 前漢 Period. What Jung refers to as the *I Ging* and Yuasa as the *Yijing* are altogether written as *The Book of Changes* in this chapter, even though it should be *Zhouyi*, since it contains *Yizhuan*. On the other hand, the original manuscript of *The Secret of the Golden Flower* is known as *Taiyi-jinhua-zongzhi* 太乙金華宗旨, and is supposed to be a compilation of the teachings of the Taoist monk named Lü Dongbin 呂洞賓 (789?–871?). Its actual year of publication was, however, much later in the Qing 清 Period. "*Liandan* 煉丹," the technology used for making the medicine called *dan* had been handed down among the Taoist monks. It consisted of *waidan* 外丹 (outer *dan*) and *neidan* 内丹 (inner *dan*). *Waidan* is external alchemy, the medicine for the exterior of the human body, and it included what we know today as Chinese herbal medicine and even acupuncture and moxibustion. *Neidan*, on the contrary, is internal alchemy, the medicine for the interior of human body, and it included breathing methods such as *qixi* 氣息 and *taixi* 胎息, meditation and the method called *daoyin* 導引, which was thought to improve the circulation by controlling the flow of *qi* 氣. *The Secret of the Golden Flower* was a book on this *Neidan.* In my research, I referred to the Japanese translation of this book by YUASA Yasuo and SADAKATA Akio 定方昭夫. This translation traces back to the sourcebook Wilhelm used, titled *Changshengshu/ Xumingfang Taiyi-jinhua-zongzhi* 長生術/續命方太乙金華宗旨 (with commentary by Zhanran Huizhenzi 湛然慧真子, Beijing,

> The former (*The Book of Changes*) became the inspiration for the views on nature Jung came up with and referred to as "synchronicity" later in his life. This suggests to us the topic for research on a new way of thinking that connects psychology and physics. More generally put, it is a subject that will allow us to reconsider the relationship between the human mind and the natural environment. *The Secret of the Golden Flower* is a book on Taoist meditation. It explores the functions of the human mind from the psychological viewpoint of the unconsciousness. While *The Book of Changes* indicates the relationship between human beings within the natural environment and space, *The Secret of the Golden Flower* deals with the internal structure of the human mind, or the "soul." (Yuasa et al. 2001, pp. 12–13)

3.2 Synchronicity: Jung and The Book of Changes

How, then, did Jung find the idea of synchronicity from *The Book of Changes*? Two topics of research are contained within this question. One is about how Jung came to know about *The Book of Changes*. The other is about the essence of the divination described in the book, which is a very complex question relating to how it is that divination can turn out to be true.

As widely known, the first encounter of Jung and Wilhelm goes back to 1920. It was the first time Jung witnessed divination being performed by Wilhelm. Jung seemed to have already known of the existence and method of divination from reading the English version of *The Book of Changes* translated by James Legge (1815–1897) (Legge 1882). This was 4 years prior to the publication of Wilhelm's *I Ging* in 1924. As is well known, the encounter turned out to be the starting point for Jung's "synchronicity" theory, also known as the "acausal combination of events," which he began to explore in the 1950s.

First of all, let us clarify the term "synchronicity." In the *Richard Wilhelm Memorial Lecture* in Munich, May 1, 1930, Jung pointed out that while translating and annotating *The Book of Changes* was the greatest achievement of Wilhelm, the basic principle for the practice of *The Book of Changes* contradicted by the Western scientific (causal) view of the world. He gave this principle a temporary name of "*synchronistisches Prinzip* (synchronicity principle)." It suggested the existence of psychologically corresponding phenomena, in which different kinds of linkages can be found when there is no causal relation. These linkages occur when multiple events happen within relative simultaneity. For example, when the same idea, symbol or state of mind simultaneously emerges in different places and, though there is no causal way to explain it, a kind of corresponding phenomenon can be identified (Yuasa and Sadakata 1980, pp. 15–18).[18] Given that this understanding is correct,

1921), and corrected areas where the text was mistranslated, while also referring to other sources such as Miyuki Mokusen 目幸黙遷 (1967). For more details, refer to the "Translator Description" by Yuasa, attached at the end of *The Secret of the Golden Flower* (1980).

[18] The subject of the research for this thesis is the philosophy of Yuasa and not Jungian psychology. Therefore I will refrain from going to detail about Jung's thoughts in order to verify Yuasa's understanding.

what Jung called a non-causal corresponding relationship, i.e., the synchronicity relationship, is, according to *The Book of Changes*, the corresponding relationship between an event that is supposed to happen in the future and the symbol shown by the divination sign (*guaxiang* 卦象: the image of deviation),[19] that cannot be explained with causality. In order to elucidate the corresponding relationship, as Jung had pointed out, it would require verifying the validity of the divination in *The Book of Changes*. However, Jung was not interested in verifying it and instead postulated its validity by believing in the words of his late friend (Wilhelm), stating that he was "only interested in the surprising fact that the "hidden characteristic" of the Zeit-moment (time sequence) has become readable when expressed through the divination in *The Book of Changes*" (ibid.). This "hidden characteristic of a time sequence" and the "non-causal corresponding relationship" are comprised of the two sides of the same coin, only substituting a time sequence for synchronicity. It means nothing much more than just that there is some sort of corresponding relationship between what is supposed to happen in the future and the result of the fortune telling. The synchronicity in *The Book of Changes*, which inspired Jung, is about a certain type of corresponding relationship that cannot be explained with causality, of which the example can be seen in the relationship between a future event and the result of fortune-telling or divination (prediction). There certainly is an unignorable difference between Jung, who presumed the validity of future predictions, and Yuasa, who tried to clarify this through body-mind theory. This issue will be discussed in detail later on.

The primary difference between the standpoints of James LEGGE and Jung (or Wilhelm, to be precise) in addressing *The Book of Changes* is another topic that needs to be dealt with. While Legge reads *The Book of Changes* as a textbook for Chinese classics, Wilhelm read it as a book of fortune telling that included the practical skill of the divination (占筮 *senzei* in Japanese or *zhanshi* in Chinese, a form of deviation which uses 50 sticks). A similar difference can be observed in their cooperators; WANG Tao 王韜 (1828–1897) who assisted Legge, and LAO Naixuan 勞乃宣 (1843–1921),[20] who gave lessons on *The Book of Changes* and *The Secret of Golden Flower* to Wilhelm. Wang Tao was raised in an environment of Confucian refinement and spent his youth aspiring to be someone who passed the Imperial

[19] In *The Book of Changes*, the symbols for *Yin* 陰 and *Yang* 陽 are both classified as *Yao* 爻, i.e. *Yin-Yao* 陰爻 or *Yang-Yao* 陽爻. A group of six *Yao* in a hexagram is referred to as *Gua* 卦(divinatory symbol). It explains all phases of the existence of the universe, or *Yuzhou* 宇宙 (*Yu* stand for space, *Zhou* for time), can be described with 64 hexagrams (*Gua*) and 384 *Yao*. According to the logic of *The Book of Changes*, the divination (占筮 *senzei* in Japanese or *zhanshi* in Chinese) utilizing fifty skinny sticks of a plant called *Shi* 蓍 will always prove to be true. No one knows for sure if it is true of not. According to Jung, it is questionable whether the creators of *The Book of Changes,* Wenwang 文王, the king of ancient China, and Zhougong 周公, one of the highest nobles of the time, (both of them are legendary figures, I assume) were onto something or not. Only experience will provide the answer to this problem. ("Commemorating Richard Wilhelm." Ibid.: 19).

[20] This is the same person as "Lau Nai Süan," whose name appears in Wilhelm's *I Ging*, *Vorrede zur ersten Ausgabe*. He is said to have passed away when the first edition of this book was printed in 1924.

examinations 科舉. He was quite an intellectual at the time. Upon Legge's request, WANG Tao assisted him in translating *The Book of Changes,* and completed it around 1870, when he was still at the prime of his life. LAO Naixuan, on the other hand, succeed in his official career through passing the Imperial examination 進士 in 1871 (the 10th year of *Tongzhi* 同治十年). He reached the positions of the general manager of the Imperial University of Peking 京師大學堂總監督 and the Deputy Minister of Education 學部副大臣 in 1911 (the 3rd year of *Xuantong* 宣統三年), and retired after the abdication of Emperor Xuantong 宣統帝. After that, Wilhelm invited LAO Naixuan as a manager of *Zunkongwenshe* 尊孔文社, which had been established in Qingdao 青島 by Wilhelm.[21] Thus, LAO Naixuan began to teach Wilhelm about *The Book of Changes*. By the time of his encounter with Wilhelm, LAO Naixuan had been reaching his own maturing period during his last years. I cannot help but overlay the images of a researcher in his prime and an elderly sage to them respectively. Though it has been overlooked, this fact has a lot to do with Jung's, and eventually Yuasa's, understanding of the thought in *The Book of Changes*.

3.3 Divining the Future and Intuition

Yuasa, who could, unlike Jung, fluently read Chinese classics and was familiar with *senzei/zhanshi* (devining the future), offered a considerably unique interpretation of the issue Jung had postulated, namely the validity of the divination in *The Book of Changes*. Even while it followed the findings of Jung, the content of it was far more in-depth. Unfortunately, to this day, this issue is treated as somewhat illogical and ridiculous by the majority of Chinese philosophy researchers of both China and Japan to contemplate in detail, and they tend to focus only on the thought or bibliographical interests.

Yuasa's understanding adopted the mind-body theory as his basic viewpoint. According to his explanation, if we were to separate the sensory organs of the body from the mind, a person could only know the state of a current space, and would be unable to know either the past or the future, because only the mind can perceive them. The past is recollected from the memory and reproduced in the mind. The future is also predicted and imagined within the mind. In that case, when we recognize the status of spatial things through our senses and thoughts (function of the mind) and infer intellectually the status of the past (e.g., the subject of anthropological studies) or the future (e.g., weather forecast) based on it, we habitually follow the principle of causality in order to make decisions. It is no different from the principle of modern science up to this point. However, *senzei/zhanshi* is obviously outside the principle of causality. According to Yuasa, in the act of *senzei/zhanshi,* even though the person who wishes to resort to divination is required to provide the

[21] ZHAO Erxun 趙爾巽 et al., *The History of Qing Dynasty Memoir 259* 清史稿列傳第二百五十九, Zhonghua Book Company, 1977: 12825.

fortune teller with some information, such as in what situation he needs to know the future or what kind of state he is currently in, it is impossible to predict the state of the future (or to foresee it) based just on empirical data. Yuasa reasoned that it is the function of "intuition" that makes foreseeing possible (Yuasa 1995, pp. 126–127). Defining strictly what he means by "intuition" is actually an extremely difficult problem to solve.

To delve into this issue, we must move away from this point for the moment in order to address the following matter. Those who are familiar with Buddhist studies may see the association between what Yuasa calls "intuition" and *prajñā*, or "undifferientiated cognition (*nirvikalpaka-jñāna*)." While *vijñāna* recognizes the object analytically, *prajñā* represents the superb cognitive capability that allows total and essential understanding without the help of analysis. Coincidentally, its Japanese translation is "*chokkan chi* 直観智 (intuitive wisdom)." Considering Yuasa's depth of understanding regarding Buddhist studies, he must have associated "intuition" with "undifferentiated cognition" in the back of his mind. However, I speculate that if we were to address "intuition" and "undifferentiated cognition" in the same context, the analysis here may move in the direction of Buddhist epistemology, and it would deviate from Yuasa's intention. It would lack the perspective that comes from the mind-body issue. It may come as unexpected to the reader, but in fact Yuasa seeks to explain "intuition" from the perspective of depth psychology.

Yuasa states that if there were a function of such "intuition," – and it was obvious to him that there was – it would not be located in sensory organs, or in the consciousness in which intellectual judgments are performed. According to him, while senses and judgments belong to the field of the consciousness, the mind includes the field of the unconsciousness, which usually does not appear on the surface of consciousness. The reason why we can recollect our past is because we have an access to distant memories stored in the field of unconsciousness. It is in the deep part of this unconsciousness, that is, the field of the mind, where the function of intuition is kept. Based on such an argument, Yuasa stated that, "the act of 'divining' is, psychologically speaking, to know the state of spatial things in the future (or the past) through the intuition that issues forth from the unconsciousness" (Yuasa 1995: 128).

At first, I was so preoccupied with the framework of conventional Chinese philosophy that I did not have the experience of viewing the essence of *senzei/zhanshi* from the standpoint of mind-body theory as Yuasa had suggested. Come to think of it, as Yuasa pointed out, *senzei/zhanshi* also is a type of *waza* 技 (technique). Which means, it will require proficiency in order to master the *waza*. Thought *waza* itself is superficially an action and body movement, but in terms of its process, it also requires the proficiency of the mind within the body. To attain proficiency of the mind, as Yuasa put it, you would need to dig into the deeper part of the mind, that is, the field of unconsciousness, through the practice of attaining proficiency of this *waza*. In the words of Wilhelm, it is to revitalize the unconsciousness.[22] That is also to sharpen the intuition. The question is, then, in what specific way can it be

[22] These are the words of Wilhelm quoted by Yuasa (Yuasa et al. 2001: 57)

achieved. My explanation of Yuasa's viewpoint suggests that the unified improvement of the mind and body is at the root of this theory. For example, in an everyday-context, polishing the craftsmanship needed for one's occupation and contemplating quietly in order to abstain from rash and blind actions exist in a front-back relation. Yoga and meditation in Buddhism, *zazen* 座禅 (in Japanese, *zuochan* 坐禅 in Chinese) meditation of the *Zen* sect, *seiza* 静坐 (in Japanese, *jingzuo* in Chinese), an ascetic practice that followed *zazen/zuochan* in Chinese Confucianism 儒教, the previously-mentioned internal alchemy of Taoism 道教, or the rigorous ascetic exercises seen in Japanese traditional Buddhism and Shintoism are all examples of unified training methods for mind and body. Also, the idea of imperturbability advocated by Mencius 孟子 in ancient China during the fourth century B.C. which aimed to keep one's mind imperturbable to external events while staying in this earthly world, must have been another training method. It seems that the more we search, the more varieties of the Eastern approach to mind-body training can be found.

Based on this understanding, it is not hard to imagine that the role of LAO Naixuan to Wilhelm was not just that of a collaborator but of a respectable *laoshi* 老師,[23] a *Lehrer* (teacher) or a sage, which was unlike WANG Tao, who was just a collaborator to Legge. As KE Shaomin 柯劭忞 (1848–1933) recorded in *The Epitaph of Lao Naixuan* 勞乃宣墓誌銘, Wilhelm learned *The Book of Changes* from LAO Naixuan. Wilhelm's knowledge of the book came from LAO Naixuan's systematic explanation. In the description of Wilhelm, LAO Naixuan was not only a scholar who was proficient in *The Book of Changes*, but also a practitioner who was well versed in divination.[24] It may not be too daring to assume that the person who was responsible for introducing Wilhelm to *senzei/zhanshi* and had him convinced of the validity of its fortune-telling ability was LAO Naixuan. Also, it is clear that divination and the internal alchemy were in complete unison for him. This surely was passed onto Wilhelm, who handed it down to Jung (he proceeded to pursue the art of alchemy thereafter). Yuasa also was a practitioner of internal alchemy. This is where I would imagine Yuasa got his inspiration. Whether based on conviction or intuition, everything is outside the law of causality. Yuasa's explanation above does not exceed the framework of Jung's synchronicity theory. However, if we could view things without the spell of causality for once, we would notice how far our life in reality is from the laws of causality. Causality does not provide an explanation for why the *Mona Lisa* is still charming or why the music of Beethoven still fascinates people around the world. According to the laws of causality, a couple of beautiful people who fall in love with each other may be understood as nothing more than two people caught up in a kind of feverish sickness (*folie à deux*). Encountering people is yet another issue. For the encounters that have no definite causal relationship, the

[23] The Chinese word "*laoshi* 老師" does not always indicate an older teacher. The Chinese character "老 *lao*" stands for the sense of respect that includes an acknowledgement of the richness of knowledge and experience that a teacher has.

[24] WANG Zhaoyong 汪兆鏞, *Beizhuanjisanbian*,vol.8, buyuan 5 碑傳集三編・卷八 部院五, (refers to: ZHOU Junfu 周駿富 et al., *The biography of the Qing Dynasty* 清代傳記叢刊・綜錄類, Taiwan: Mingwen Book Company, 1985: 124–435).

Japanese language uses the term "*en* 縁," which has its origin in Buddhism. Chinese also uses a similar term "*yuanfen* 緣分." The use of *senzei/zhanshi* is requested within this reality of people's everyday daily lives, so it is, in fact, reflecting the real state of people's actual lives. There are many methods of divination around the world that are utilized by people, including astrology and tarot cards. If there is a statistical comparison available, I would like to know, in the United States – world leader in space exploration – which receives more expenditure astrology or astronomy. By contrast, it is obviously that there is no need to use divination about sending a man to Mars.

3.4 Ethical or Moral Cultivation

The interpretation Yuasa offered can be summed up by stating that *senzei/zhanshi* is naturally affected by what could be called the personality factors (the degree of ethical/moral cultivation) of the people concerned. Jung did not mention anything regarding this point; neither did Yuasa discuss this topic to any great extent. However, from my point of view, Yuasa's interpretation of *The Book of Changes* goes far beyond Jung's understanding on this point. The sources to support his thesis can be found in the twenty-some articles on *senzei/zhanshi* listed in some of the documentary records of ancient China, such as the *Guoyu* 國語 (means Monarchs' Remarks) and the *Chunqiu-zuoshizhuan* 春秋左氏傳.[25] Two examples from them shall be addressed hereafter.

First, I would like to introduce the cases of *senzei/zhanshi* recorded in history from 564 B.C. (9th year of Duke Xiang 襄公's rule. MU Jiang 穆姜, the mother of Duke Cheng 成公 of the land of Lu 魯, had committed adultery with SHUSUN Qiaoru 叔孫僑如, one of the three Court nobles and proposed usurpation of the regime, supporting Prince Yan 偃, out of the many princes. However, the scheme failed and Qiaoru escaped to the State of Qi 齊, while Mu Jiang was imprisoned in the East Palace. This incident took place in 575 B.C., which means she was imprisoned for 11 years until her death in 564 B.C. At the beginning of her imprisonment, the result of the divination done by the official diviner was "hexagram *Gen* 艮 to

[25] *The Spring and Autumn Annals* (Chinese: 春秋; pinyin: *Chunqiu*) is an historical text that chronicles the State of Lu 魯, a region in ancient China where Confucius was born. *The Zuo's Tradition: Commentary on the Spring and Autumn Annals* (Chinese: 春秋左氏傳; pinyin: *Chunqiu Zuoshizhuan*) is believed to be written by Confucius, with the addition of a commentary or *Zhuan* 傳 written by ZUO Qiuming 左丘明; however, there are no documents that can prove this legend. A more realistic assessment would be to assume that it was created by a group of Confucians during the *Zhanguo* 戰國 or Warring State Period. *The Discourses of the State* (Chinese: 國語; pinyin: *Guoyu*) is another written work attributed to ZUO Qiuming, but with no solid proof to back it up. It could also have been written in the *Zhanguo* Period.

I have offered a detailed analysis of these articles about *senzei/zhanshi* in my book *The Development of the Heaven's Will in Ancient China—Ancient Confucian Philosophy and the Logic of Yijing*. It will not be dealt with here due to the limited space.

hexagram *Sui* 隨." The original hexagram *Gen*, the explicit sign, which indicated the current state, then would be changed to hexagram *Sui*, the implicit sign, which indicated the direction things are heading. The hexagram *Gen* stands for "to remain," which suggested bad luck in this situation. However, the official diviner seemed to have followed the commentary of the hexagram *Sui* and consoled MU Jiang telling her that her divination was good and that she would be released. But MU Jiang refused to accept it. She explained the reason for her refusal, stating that the only situation in which the hexagram *Sui* "does not contain punishment" is when it was given to a virtuous person. Therefore, unless she gets virtue herself, there is no use in receiving the hexagram Sui because it will certainly contain some punishment. MU Jiang's refusal of the fortune told by the divination officer, conclusively indicates the fact that the divination had already been transformed into a symbolic system by this time. By "symbolic," I mean that there was thought to be no inextricable order of priority between divination symbols, their commentary and divination generated from them. The attitude of MU Jiang was something that kept her away from effacing herself in the midst of fatalism. It was more parallel to the humanistic way of thinking that was emerging from the *Chunqiu* to Warring State periods. Confucius 孔子 (551–479 B.C.) appeared in the middle of this trend toward humanism.

Yuasa did not overlook this point. He stated that the "time factor in the divination is an issue related to the specific decisions made by people at certain points in time, people who are living their lives and experiencing each specific circumstance that accompanies each moment in time. This is, different from a scientific forecast. The divination is not established on the basis of causalistic determinism. Instead, it leaves the decisions to the people's free will. In other words, issues of both historicity and morality are brought into the process." (Yuasa 1995: 136).

The next example is of the divination regarding the rebellion of NAN Kuai 南蒯 recorded in the history of 530 B.C. (12th year of Duke Zhao 昭公's rule). This episode has a preceding story, and is related to the dispute around the initiative of *Sanqing* 三卿 (three noble families) that practically governed the regime of the state of Lu 鲁. A summary of the event is as follows: NAN Kuai was the supervisor of the Fei 費 town, which was a territory of JI Pingzi 季平子. NAN Kuai was dissatisfied with his master JI Pingzi because he was not receiving the respect he deserved based on his job. He plotted to expel JI Pingzi along with the younger brother of Duke Zhao, Prince Yin 慭. NAN Kuai performed *senzei/zhanshi* prior to raising his rebellious army, and received the divination sign of "from hexagram *Kun* 坤 to hexagram *Bi* 比." The commentary of the divination "*Kun*" six-five[26] suggested a "great luck with a yellow skirt." Yellow was the color of the center and symbolized the king, whereas the skirt was something that covered the lower body. Therefore, this divination could be interpreted as the king being humbled in some way, which would mean a great luck would come to NAN Kuai. Though that was his impression, just to make sure, he consulted the high steward of the state of Lu, ZIFU Huibo 子服惠伯 for his

[26]Technically speaking, *The Book of Changes* uses numbers 六 (six) and 九 (nine) to symbolize *yin yao* 陰爻and *yang yao* 陽爻, respectively. The characters 六五 (six-five) indicate the fifth *yin yao* from the bottom. Each *yao* has words that symbolize them, which are called "*yaoci* 爻辞."

interpretation. The record shows that Huibo answered that according to what he had learned, when a man is to make an action, his interior and exterior must be in unison (sincerity), and he must execute it with faith (deference). He must cultivate the three virtues of sincerity, faith and deference and if he is without those three virtues, it would be unclear if the sign of "great luck" would become manifest as its divination. Moreover, the divination must not be used in divining such ominous matter as the success following a rebellion. The characteristics of the thoughts given above are to be found in how they connected the act of divining and the resulting divination using the virtue of the diviner (or, to be more precise, the person who requested the divination) as a tying medium. The idea of requiring the act of divining to be ethical in nature is emphasized in this case. As could be expected, NAN Kuai Rebellion ended in failure.

However, divination by its nature was in fact supposed to be nothing but a text for divination. Its scope should not be limited to the ethical-moral field. The interpretation indicated in *Chunqiu-zuoshizhuan*, which brought a sense of morality into the divination, concurrently bestowed a new interpretation to it. Following the Former-Han period, Confucianism integrated divination within its system and by doing so obtained a theory of time and space, which had not been originally subsumed within it. At the same time, *The Book of Changes* established its authority as the Sacred Book, enlisting its name on the top of the Five Sacred Books. Considering these facts, the thought structure in the *Chunqiu-zuoshizhuan,* which integrated the divination and a sense of morality and ethics, has an important significance in terms of the development of Confucianism.

In that case, one could also view this as allowing virtuous people to disregard the use of *senzei/zhanshi*. This issue is clearly addressed in the words of Xunzi 荀子 (Master Xun) from around fourth century B.C., after Mencius, who stated that "the one who knows divination well does not divine"[27] (*Xunzi* 《荀子》, *Chapter Dalüe* 大略). However, we must pay attention to the fact that according to the logic of *The Book of Changes*, the ultimate destination for "the one who knows divination well" is that of the "sage 聖人." This fact is emphasized in the *Ten Wings* 十翼, especially in *Xicizhuan* 繫辞傳. Undoubtedly, sages do not need to divine. Sages are personifications of the Way of Heaven 天道/天, and since they exist in the dimension of Heaven 天, they can know the fate of things without divining. Though there is no discontinuation between Heaven and people in Eastern thought, they do reside in different levels. To truly know the divination is ultimately an ideal. For individual people in reality, to become familiar with the technique of the divination as they cultivated their morals and ethics, and to be able to utilize it without a mistake (since "wisdom" and "technique" are inseparable) was a process, or a staircase to becoming a sage, that is, an ideal character or personality. To put in another way, the divination exists in the gap as a medium between someone "knowing divination"

---

[27] The original sentence is "善為易者, 不占 *Shanweiyizhe, buzhan*." I have translated the character "為 *wei*" as "to study and master," which is represented as the word "知 know" in the modern language. The character "知" contains such nuances as to "administer," "govern" or to "comprehend," which goes beyond just the theories or knowledge in both Chinese and Japanese language.

while in the process of personal cultivation using ascetic moral practices, and actually being a sage who truly knows divination. According to the logic of *The Book of Changes*, it is in this space that we can find the significance of the existence for the divination, and therefore its necessity in the real world cannot be disregarded. Looking back from the viewpoint of Chinese philosophy, Yuasa's "intuition" took a step deeper into the issues related to the ethical-morality of people involved in the divination. To supplement with the words of Yuasa, "The present is the state in which consciousness and the body are connected, and the past and the future are latent in the unconsciousness that exists at the bottom of this" (Yuasa 1995: 130). Yuasa also stated that "the divination is a recognition based on unconscious intuition" (Ibid.: 101). To put it another way, the technique of the divination can be comprehended as a series of actions (*prāxis*) that shows the intuition within the unconsciousness, expressed as the divination symbols through the divining sticks.

The statement above regarding *The Book of Changes* may seem to deviate from the main theme of this thesis. But it is my attempt to amplify some of the main issues Yuasa addresses, such as "divination" and "intuition," from the viewpoint of Chinese philosophy. As already mentioned, Yuasa states that his interest in the Eastern thought was renewed by Jung's interpretation of *The Book of Changes*. Jung had taken a pioneering role in outlining the "movement to rediscover the meaning and values of the spiritual legacy of the West, which modern rationalism had lost track of" (Yuasa et al. 2001, p. 29) from a novel point of view that was new to the history of Western thought.

4 Conclusion

Keeping this in mind, I suspect that, even though Yuasa's philosophical interest was facing the same direction as Jung's eyes were (even though he had no interest in astrology), at the base it is built more on ethics than on psychology. More precisely speaking, as will be mentioned later on, what we see here is the connection of Watsuji Tetsurō's ethical system and Jungian psychology. To put it still another way, Yuasa's philosophy on its own can be considered an integration of the wisdom of the East and West. Though an outsider to psychology, I have always sensed the deep infusion of ethical motifs in the "(mainly Jungian) psychology" Yuasa often discusses. This tendency is indicated in the following passage:

> For example, meditation methods are one of the "*waza*" used to train the "mind." Though originally *qigong*/*kikō* 氣功 garnered attention as a method for enhancing the health of the body, it has its historical root in ascetic practice. Ascetic practices originally had the ethical goal of achieving an improvement in the spirituality of a human being, that is, an improvement of the "personality," through training the abilities of mind and body, and bringing out that person's inherent ability. Psychologists of today suggest "self-actualization" or the establishment of "identity" as goals necessary for contemporary people. It is often said that the young generation of today has lost their identity. Put simply, it means they are in a state in which they have lost the meaning and value in living their lives. Nowadays it seems that this psychological feeling has spread throughout the entire society.

> Jung called the purpose of psychotherapy "individuation." That may be a somewhat difficult expression, but it means "to become a true individual." We tend to think that we are "individuals," but the truth is, we do not grasp our own inherent ability, or what to do with our lives. Jung advocated the need to discover and cultivate it. (Ibid.: 25–26)

Yuasa interpreted Jung's "individuation" as meaning "to become a true individual." He also put it in his own words and described it as the "establishment of identity." Moreover, he states that this process of focusing on one's inner self, or so to speak, the process of self-awakening is what leads to our ethical goals as human beings, that is, the improvement of "personality." This at a glance may look no different from the Western ethical framework that emerged from the modern period, but in fact it strikes me as expressing a completely opposite perspective.

What needs to be paid special attention to here is this "becoming an individual." To put it another way, it is to reach the state of "having a well-balanced, normal sense of ethics" (Ibid.: 59). A human being is something to become. As opposed to this, the modern Western view of human beings sets the abstract individual as the philosophical subject at the center of contemplation. It is as if to suppose that an intellectual self-consciousness is the essence of being a human. Descartes' idea of *cogito* is precisely an example of this idea. With Descartes' idea in mind, Yuasa points out that "this standpoint locates the essence of being human into the intellectual and rational part of the "mind," which is merely a part of being human, and concurrently ignores the body." (Ibid.: 61). According to Yuasa, "in the basic attitude of the modern Western thought, philosophy (originated in the modern West) excluded embodiment, and science eliminated spirituality from being human" (ibid.).

In such a context, Yuasa suggests that the most basic skeletal structure needed when thinking about ethics is not the "first person (I)," but rather the "second person relationship." He describes this viewpoint in his own words, saying "(For example, just as a marital relationship represents this pattern), it is a relationship where the self and the other know each other, and the body and mind are perceived as one" (Ibid.) (a supplementary explanation will be provided for this rather complicated theory later on). Moreover, he also refers to this two-person relationship, using "*aidagara* 間柄," a term reinvented by his teacher WATSUJI Tetsurō. According to the common understanding, I would like to suggest defining this word as a thought that perceives individuals and society as something that cannot exist independently, but only in mutual dependency. Yuasa agreed with the argument advocated in the ethics of Watsuji, which considered the establishing of the individual as the starting point for contemplating ethics as the major mistake evident in the modern West's view of human nature. Yuasa made the following in support of this issue:

> Watsuji's thinking, in which the existence of "*aidagara*" preceded that of the individual, included issues regarding "sexuality." His various discussions in *Ethics* are about community, and what comes first is the "two-person community," namely a marital relationship. Human existence first starts in this world based on the sexual relationship between a man and a woman. Therefore, we must recognize that its existence starts in "*aidagara.*" According to Watsuji's idea, the first-person and third-person relationships are only abstract forms of human existence derived from "*aidagara*" as a second-person relationship. (Ibid.)

As Yuasa mentions, Watsuji starts the third chapter of his *Ethics* entitled "Ethical Organization" with an explanation of the two-person community (marital relationship). In terms of context and meaning, Yuasa's explanation strikes me as accurate and I would have no argument with it. What Watsuji describes as the mutual engagement of mind and body between a married couple is what Yuasa calls the unity of mind and body between a married couple. I would just say that my impression is that Yuasa puts greater emphasis on the perspective coming from the body-mind theory. In the paragraph that follows the quote, Yuasa gives an example from Husserl to show how modern Western thought, which is grounded in individualism, tried to derive the human relationships as second-person relationships that issue forth from first-person self-consciousness. He indicates that the reason why Husserl failed to logically explain the existence of another's "personality" from the viewpoint of self-consciousness was due to an error in the way the problem itself was set up. Yuasa is consistent with Watsuji on this point.

However, attention needs to be paid to what Yuasa expressed at the end of these statements, saying, "'*aidagara*" precedes intellectual logic and thus already dominates us psychologically. Therefore, issues related to ethics must never be discussed without addressing psychology" (Ibid.: 62). This assertion could be interpreted as an indication by Yuasa that Watsuji's ethics was missing the psychological perspective. It is not the intention of this thesis to step into and analyze Watsuji's enormous ethical system, nor would it capable of doing so. However, in order to understand Yuasa's intention, this basic characteristic cannot be left unmentioned.

According to the "Commentary" offered by the successor of Watsuji, Professor KANEKO Takezō, Watsuji's particular writing *Ethics* was completed between 1934 and 1949 (Watsuji 1962).[28] The gist of his explanation suggests that, judging from drafts of *Criticism of Homeros* (1946), *The Importance of Primitive Christianity in Cultural History* (1926) and *Practical Philosophy of Primitive Buddhism* (1927), Watsuji was coming close to completing his systematization of ethics. What "facilitated the crystallization" of it was his one-year visit to Europe from 1927. Watsuji was deeply impressed by Heidegger's *Sein und Zeit* (1927), and at the same time could not help but feel repelled by it. Kaneko continues on by summarizing the preface of *Climate and Culture*, which describes Watsuji's impression of reading *Being and Time*. It was intriguing that Heidegger was able to grasp the structure of human existence from time, but the question remains on why spatiality was not, at the same time, utilized as a fundamental aspect of the structure of human existence.[29] Temporality and spatiality are united and inseparable. The reason why Heidegger stayed with just temporality was because what he called "*Dasein*" ("there-being" in Heidegger) focused solely on the individual. Human existence should be comprehended through the dual structure of the individual and society.

---

[28] I was given many suggestions from this excellent "commentary" about the following statements on Watsuji's ethics.

[29] Watsuji also recognizes that spatiality is not completely absent in Heidegger's argument. Watsuji states that, "however, it has lost most of its shadow in the strong illumination of temporality. That is where I sensed the limit of Heidegger's work" (Watsuji 1935: 4).

Had that been done, temporality and spatiality would have been united. More specifically, temporality and spatiality should have been united as historicity and endemicity (*fūdosei* 風土性, surroundings related with climatic circumstances), respectively (Ibid.: 452–455).

If I were to summarize in general terms the character of Watsuji's ethics, I would say that it is an inquiry into the way (or principles) of existing in this world, in which human existence has a dual structure, comprised of both an individual aspect and a social aspect, and which is based on an ontology that takes as its defining axes temporality and spatiality (or, more specifically, historicity and endemicity). I assume that what bothered Yuasa was Watsuji's comprehension of spatiality. Strictly speaking, his direct concern was on the issue of the body that occupies the space, and more importantly, the issue of the mind that is inseparable from the body. Yuasa's true intention in stating "the issues of ethics must never be discussed without addressing psychology" was quite possibly associated with this point.

Yuasa bequeathed to us an enormous number of works that cover many conventional disciplines. He did not set forth any coherent system like Watsuji did with his *Ethics*. However, if Yuasa had not died of a sudden accident, we could have witnessed his theme of integrating the wisdom of the East and West in a more articulated and systematized form. It would undoubtedly have been a fusion of Aristotelian *μετα-φυσικά* or metaphysics (beyond the physical) and what Yuasa refers to as *μετα-ψυχικά* or meta-psychika (beyond the spiritual, a term coined by Yuasa himself), or the metaphysics of the Eastern thought. (Yuasa 1989a, pp. 205–206)[30] For instance, Ray KURZWEIL advocates that the concept of technological singularity is to be aware of the limits of human ability beyond physics. If so, Yuasa's works represent an argument in the direction of viewing technology beyond one's sight is only possible through the technique of training the human spirit and this cannot be neglected. What Yuasa referred to as "beyond the physical" and "beyond the spiritual" and his argument rising from their interactivity, at least for us who are living in massive technological society, are unmistakably important issues that should be taken seriously.

Finally, as a postscript, I'd like to mention the following. Yuasa gave me many of his works as gifts, and though I could well have been taking up a study of Yuasa with gratitude, I have been satisfied with merely being one of his readers. Studying Yuasa as a subject of research has been too profound and enormous an arena for me to step into. The more I reviewed his books, recollected the memories of his lectures and his voice, the more remorse I felt for learning so little from him. It was only after I finished reading his books related to the subject of this thesis that I finally felt as if

[30] In order to understand the process through which this concept was established, I referred to WATANABE Manabu渡邉学's essay "Searching for meta-psychika" (2015). Professor Watanabe of Nanzan University was one of Yuasa's senior graduate students. Additionally, regarding this issue, it is worth noting a statement Yuasa left with us: "When considering the movements within the contemporary world, the questions presented by Jung have not been answered clearly at all, whether from the viewpoints of philosophical history or scholarship. He still is a prophetic scholar-philosopher, who continues to pose questions to us even today" (*Science, Humanity and Ethics*, vol. 2; *Human Body Science*, vol. 11, issue 2, October 2002; *Complete Works*, vol. 17, BNP, 2012: 500).

I had at last a grasp of Yuasa's philosophy. My pen still refused to proceed because I kept on being tormented from the anxiety of failing to understand the true intention of Yuasa's philosophy.

Acknowledgements Without the enthusiastic support of Professor MURATA Kiyoshi 村田潔 of Meiji University and Dr. Thomas Taro LENNERFORS of Uppsala University, as well as Professor NAKADA Makoto 仲田誠 from the University of Tsukuba, I would have continued simply as a reader of Yuasa's philosophy. It has been my pleasure to work with the best collaborators, and I would like to express my deepest gratitude to each one of them.

References

Heidegger, M. (1927). *Sein und Zeit [Existence and Time]*. Jahrbuch für Philosophie.
Kasulis, T. (2002). *Intimacy or integrity: Philosophy and cultural difference*. Honolulu: University of Hawaii Press.
Legge, J. (1882). The Sacred Books of China, The Text of Confucianism, part II, The Yi King,. (J. Legge, Trans.) Oxford at the Clarendon Press.
Lévi-Strauss, C. (1962). *La pensée sauvage [The Savage Mind]* (Vol. 29 ed.). Paris: Plon.
Mokusen, M. (1967). *The Secret of the Golden Flower, Studies and Translation* (Dimploma Thesis ed.). C.G. Jung Institute.
Rorty, R. (1979). *Philosophy and the Mirror of Nature*. Princeton: Princeton University Press.
Rorty, R. (1982). *Consequences of Pragmatism*. Minneapolis: University of Minnesota Press.
Sartre, J.-P. (1960). *Critique de la raison dialectique: Tome I, Théorie des ensembles pratiques [Critique of Dialectical Reason]*. Paris: Gallimard.
Watanabe, M. (2015). Searching for Meta-Psychika. In *Wisdom of the Body: Inheritance and Development of the Philosophy of Yuasa*. Tokyo: BNP.
Watsuji, T. (1935). *Climate*. Tokyo: Iwanami Shoten.
Watsuji, T. (1962). *Watsuji Tetsurō Zenshu [Complete Works of Tetsurō Watsuji]* (Vol.11 ed.). Tokyo: Iwanami Shoten.
Wilhelm, R. (1924). *Das Buch der Wandlugen (I-Ging)*. Diederichs Jena.
Wilhelm, R. (1929). *Das Geheimnis der goldenen Blüte: ein chinesisches Lebensbuch*. Dornverlag.
Yuasa, Y. (1978). *Yungu to Kiristokyō [Jung and Christianity]*. Kyoto: Jinbun shoin.
Yuasa, Y. (1979a). Eastern Religions and Depth Psychology.
Yuasa, Y. (1979b). *Jung and the European Spirit*. Kyoto: Jinbun Shoin.
Yuasa, Y. (1982). *The Deeper Layers of Eastern Cultures*. Tokyo: Meicho Kankokai.
Yuasa, Y. (1984). Literary Fragments on Views of Nature and Self-Cultivation Theories in Esoteric Buddhism.
Yuasa, Y. (1987a). *A Study on Synchronicity*. Tokyo: Sannou Shuppan.
Yuasa, Y. (1987b). *The Body: Toward an Eastern Mind-Body Theory*. (T. P. Kasulis, Ed., & K. T. Nagatomo S, Trans.) Albany: State University of New York Press.
Yuasa, Y. (1989a). *Jung and the East*. Kyoto: Jinbun Shoin.
Yuasa, Y. (1989b). Knowledge in the Age of the End of Philosophy. *Japanology, 13*, 116-161.
Yuasa, Y. (1990). *Qi and Human Science*. Tokyo: Hirakawa Shuppan.
Yuasa, Y. (1991). *Ki towa nanika [What is Qi?]*. Tokyo: NHK Books.
Yuasa, Y. (1993). The Rebirth of Philosophy. In *Shūkyō to kagaku no aida [Between Religion and Science]. Tokyo: Meicho*. Tokyo: Meicho Kankō-kai.
Yuasa, Y. (1994). *Shintai no Uchèsei: Toyō to Seiyō [Cosmic View of Body: East and West]*. Tokyo: Iwanami Shoten.
Yuasa, Y. (1995). *Kyōjisei no Uchūkan: Jikan, Seimei, Shizen [Synchronicity of the Vision of the Universe: Time, Life and Nature]*. Kyoto: Jinbun Shoin.

Yuasa, Y. (1999–2003). *Complete Works of Yasuo Yuasa* (Vols. 1–17). Tokyo: Hakua Shobō and Being Net Press.

Yuasa, Y., & Sadakata, A. (1980). *Ōgon no hana no himitsu [The Secret of the Golden Flower].* Kyoto: Jinbun Shoin.

Yuasa, Y., Takahashi, Y., Ando, O., & Tanaka, K. (2001). Jungian Psychology's Inquiry into the Spirit of Modern Man – An Effort to Unify Eastern and Western Thought. In *Jungian Psychology and Modern Crisis.* Tokyo: Kawade Shobo Shinsha.

Part II
Japanese Ethics in a Technological World

Chapter 7
Japanese Traditional Vocational Ethics: Relevance and Meaning for the ICT-dependent Society

MURATA Kiyoshi

1 Introduction

Japan is a global leader in the development and usage of information and communication technology (ICT) and ICT-based information systems. The heavy chemical industry, which was a primary driver of Japan's rapid economic growth after the Second World War, suffered a downturn due to domestic market maturity and increasing competition in the globalised economy. This downturn forced Japan to adopt a knowledge-based economic structure. By the mid-1990s, ICT was recognised as a necessary strategic area to maintain Japan's competitive strength internationally. The Advanced Information and Telecommunications Society Promotion Headquarters was established in 1994, and its successor, the Strategic Headquarters for the Promotion of an Advanced Information and Telecommunications Network Society (IT Strategic Headquarters), was established within the Cabinet in 2000. The purpose of these organisations was to develop a policy environment in which all Japanese citizens could share the benefits of ICT, and to make Japan the most advanced nation in terms of ICT innovations. A series of *e-Japan Strategies* was developed by the IT Strategic Headquarters, resulting in a sophisticated broadband network infrastructure that offered cost-effective, continuous internet connections throughout the country. In 2004, the Ministry of Internal Affairs and Communications adopted the *u-Japan Policy*, which was intended to create a ubiquitous network society in which seamless network access would be achieved through the

The original version of this chapter was revised: MS Mincho font was updated throughout the book for Japanese and Chinese characters. The correction to this book is available at https://doi.org/10.1007/978-3-319-59027-1_13

MURATA Kiyoshi (✉)
School of Commerce and Centre for Business Information Ethics, Meiji University, Chiyoda, Tokyo, Japan
e-mail: kmurata@meiji.ac.jp

Thomas Taro LENNERFORS, MURATA Kiyoshi (eds.), *Tetsugaku Companion to Japanese Ethics and Technology*, Tetsugaku Companions to Japanese Philosophy 1,
https://doi.org/10.1007/978-3-319-59027-1_7

integration of wired and wireless networks, by means of which people could benefit from the high-quality services in a secure and safe ICT environment. Multifunctional mobile (feature) phones, which have unique features compared to those used in other countries, were widely used in Japan as handheld information terminals from the late 1990s, and were replaced by smartphones when the Apple iPhone 3G became available on the Japanese market in August 2008.

However, issues regarding information ethics have not been appropriately addressed, even though Japan has been increasing its dependence on ICT. Although some point to recent increases in the Japanese public's awareness of information privacy rights, it has increased as a result of shallow news reports on personal information leaks and the risks of identity theft (Orito et al. 2008); effective measures for protecting these rights have not been taken. The 2005 Act on Protection of Private Information (APPI; Act No. 57 of 2003) has not served as a well-functioning basis for the protection and proper use of personal data (Murata and Orito 2008; Orito and Murata 2008). This act, as well as the revised APPI which passed in September 2015, is a characteristically pro-economic law, not a pro-privacy law, emphasising the importance of protecting information privacy without mentioning other dimensions of privacy such as bodily and decisional privacy (Murata and Orito 2016). There are no legal regulations on CCTV cameras in Japan, and ICT-based surveillance issues such as behavioural targeting techniques employed by online advertisers and state surveillance practices revealed by Edward SNOWDEN have not attracted popular interest, partly because the Japanese mass media rarely report on such issues (Murata et al. 2017).

Morally questionable ICT-based services, such as mobile dating sites that juveniles can easily access, continue to be set up by businesses. Furthermore, effective measures have not been taken against the online distribution of child pornography, though the Revised Child Pornography Law (Act No. 79 of 2014) makes the production, sale, distribution and possession of child pornography illegal. The production and sale of video games containing violence and sexual content remain effectively unregulated. The infringement of intellectual property rights through the use of peer-to-peer file sharing systems such as Winny, Share and their successors remains substantially unregulated even though copyright laws have been revised to forbid individuals from downloading copyrighted online content for their private use. Professional and gender issues related to ICT have received little or no attention from the Japanese people.

Considering the global nature of ICT and ICT-based information systems, and the fact that Japan is one of the leading nations in ICT, poor responses to these issues concerning information ethics may harm social welfare not only in Japan but also abroad. To appropriately address such crucial concerns, cultivating an attitude of professionalism among ICT engineers should be recognised as an urgent priority, as they play a key role in dealing with this and other pressing issues. However, there are serious obstacles to realising that cultivation.

First, the social importance of ICT professionalism is not acknowledged in Japan. There is no well-organised educational programme for ICT engineering students at Japanese universities to cultivate an attitude of professionalism. Generalised ethical education programmes for engineering students were developed based on a Western-model and are provided to students regardless of their majors. Few learning opportunities for developing and/or maintaining a professional outlook is given to ICT workers who did not necessarily receive an engineering education, despite the fact that many of them work in highly stressful and physically demanding environments that can disrupt their levels of professionalism (Murata 2013).

A more serious obstacle is that Japan's traditional forms of ethical thinking and discourse have almost died out. In addition, the Japanese have lost sight of their core vocational ethical values, which were commonly recognised in the past, through the process of developing a capitalist democracy, especially after the Second World War. This tendency has accelerated since the adoption of neoliberal policies by the Koizumi Administration in the early part of this century. Considering the local nature of human beings, in contrast with the global nature of ICT, any policy addressing ICT-related ethical issues will be ineffective if it is not supported by the ethical values shared by those affected by its implementation. Therefore, renewing traditional forms of ethical thinking and discourse related to professional life and implementing necessary revisions suitable to the current globalised, multicultural and ICT-dependent world should be recognised as a very important first step toward the development of a spirit and an attitude of professionalism among ICT engineers.

To make this first step, this chapter examines and reflects upon the historical circumstances surrounding the formation of Japanese core vocational ethical values. In particular, the ethics of the Tokugawa Era or the Edo Period (1603–1868) will be scrutinised. In that exceptionally peaceful period in Japanese history, learning and culture were highly developed, and vocational ethics were established and disseminated throughout every social stratum. These values were based on the ancient traditions of Japanese religion, which were developed through the integration or syncretisation of Shintō, Buddhism, Confucianism and Daoism. The vocational ethics developed in this era functioned as the spiritual foundation for the Meiji Restoration and the economic recovery after the Second World War, and have continued to influence Japanese workers' ways of thinking and behavioural patterns, at least on a psychological level, to the present day.

The structure of the remainder of this chapter is as follows. The next section explores the characteristics of Japanese Buddhism as the basis of Japanese vocational ethics and gives an introduction to the labour philosophy and vocational ethics advocated by SUZUKI Shōsan. In the following section, the vocational ethics of samurai, merchants and farmers developed in the Tokugawa Era are investigated in relation to their religious backgrounds. The last section examines the necessity and the possible ways of restoring and revising the Japanese core vocational ethical values in the current ICT-dependent society.

2 Religious Basis of the Formation of Professional Ethos in Japan

2.1 De-mortification of Buddhism and Gudō-ism

Sinicised Mahayana Buddhism[1] from Korea (*Baekje*) was introduced to Japan in the early to mid-sixth century as the first non-natural religion[2] in the country that had systematic doctrines.[3] After violent conflicts with the existing political force supporting primal Shintō,[4] Buddhism was officially embraced in Japan in the later part of the sixth century, and was practiced by the imperial family, as well as those nobles and local ruling families who supported the imperial court. Buddhism gradually spread, mainly, among nobles and people living in the capital towns during the Nara (710–794) and Heian (794–1185) Periods. Buddhism remained an aristocratic religion in these periods during which the Tendai and Shingon schools, the two major sects of Buddhism in Japan, were founded by Saichō 最澄 (in 806) and Kūkai 空海 (in 816), both of whom had lived in China (during the Tang Dynasty) as scholar monks. Heian Buddhism, which incorporated the teachings of esoteric Buddhism, was acknowledged by the imperial family and the nobles as a measure to acquire and maintain power, as well as a way to protect the nation from the threat of plagues and disasters by the help of Buddha.[5]

From the end of the Heian Period to the middle of the Kamakura Period (1185–1333), the religious task of offering the general public some means of achieving a sense of salvation or enlightenment was addressed. Samurai or *bushi* 武士, for example, whose duties included sinful homicide, suffered the stigma of blood pollution in a society based on Shintō concepts of *kegare* 穢れ (pollution) and *goryō* 御霊 (vengeful departed soul) (Ikegami 1995). Those associated with food production and farming suffered from a maelstrom of wars and famines and an expanding disparity in wealth. Buddhist priests faced the necessity of addressing these issues. New Buddhist sects of so-called Kamakura Buddhism, such as the Pure Land or Jōdo sects (Jōdo shū, Jōdo Shinshū and Jishū), the Zen sects (Rinzai shū and Sōtō shū) and Nichiren Buddhism (Hokke shū), were thus founded by monks who had studied Tendai Buddhism. One of the common characteristics of these Japanised

[1] Original Buddhism probably first came to China (Han Dynasty) in the first century, and started to be syncretised with the indigenous religions including Confucianism and Daoism during the third century at the latest.

[2] Differently from Christianity or Islam, Buddhism is not a revealed religion but a religion proposed by a person (Gautama Buddha).

[3] Confucianism was also introduced from *Baekje*. In the early six-century, Five Classics scholars were sent to Japan from that country (Bito 2000).

[4] Shintō 神道 is the Japanese traditional natural religion developed based on the ancient traditions of nature worship, animism and shamanism, which did not have any systematic doctrine or scripture at the time. The theologies and theoretical treatises of this religion were thereafter developed using concepts, ideas, terms and logical systems of Buddhism and Confucianism (Inoue 1998).

[5] This typically shows the Japanese tendency to turn to religion for realising worldly benefits.

Buddhist sects was the concept of de-mortification, or the idea that people could achieve enlightenment and spiritual salvation not through rigorous learning and ascetic training, with strict commandments, in a Buddhist monastery, but through relatively easy religious practices such as giving Buddhist invocations (*nenbutsu* 念仏) or Zen sitting meditation (*zazen* 坐禅)[6] in normal secular life (Teranishi 2014).

De-mortification was introduced by the *hongaku* 本覚 (true awareness) philosophy in the Tendai sect. This philosophy was based on the thought that real-world objects and events were actually the truth of Buddha (*jijōjū* 事常住), and insisted that everyone was born with Buddha nature (*issai shujō shitsu'u busshō* 一切衆生悉有仏性) and even *bonpu* 凡夫, implying that people who did not understand the Buddhist truths and were harassed by earthly desires, could still become Buddha themselves (Bito 2000; Teranishi 2014). The *hongaku* philosophy turned a dualistic world-view composed of denial of life on earth and desire for well-being in the afterlife that had been formed under the influence of the prevalent tenets of Nara and Heian Buddhism[7] into a positive perspective of the world (Murakami, et al. 1979; Teranishi 2014), and cultivated an understanding of the oneness of life and death (*shōji funi* 生死不二) (Ikegami 1995).

Teranishi (2014) pointed out that the characteristics of Japanese economic behaviour and economic systems have their origin in the Japanese idea of *gudō*-ism (*gudō shugi* 求道主義) or seeking the truth or the Way (*michi* 道) in occupational (especially artistic) activities to achieve enlightenment or to gain a religious view of the world and humanity in normal secular life. He claimed that this attitude had been formed over many generations and was based on the de-mortification of Japanese Buddhism and the resultant liberation of the general public from rigorous Buddhist practices to achieve salvation. The degree of a person's *gudō* (truth seeking) attainment was evaluated by others who were considered to be involved or could benefit from his/her occupational activities, not by Buddhist priests, Buddha or god. Such an attitude of *gudō*-ism and its evaluation scheme led to the traditions of: (a) Japanese individualism, whereby a personal objective of receiving a high valuation from relevant parties on the degree of one's self-actualisation in terms of the achievement of enlightenment in one's secular occupational life, as opposed to competing with others,[8] was emphasised;[9] and (b) an economic system centred on demand-driven production where activities for producing quality goods and services were conducted in response to customers' demands and the attainment level of a producer's occupational *gudō* was evaluated by consumers, while travelling merchants acted as intermediaries between the two sides. In addition, the development of the *gudō*-ist occupational consciousness created an awareness of the importance of accumulating human capital related to professional knowledge and skills,

[6] The Pure Land sects criticised zazen as mortification.

[7] The tenets included the concepts of Hell and the decline of Buddhism (*mappō shisō* 末法思想).

[8] The de-mortified Buddhism aimed to realise the spiritual salvation of everyone. In this doctrine, there were no competitors for anyone to defeat (Teranishi 2014).

[9] This does not mean that others' objectives or values are prioritised over a person's personal ones. Rather, a person exploits others to attain his/her objective of self-actualisation (Teranishi 2014).

and the development of *ie* 家, or systems, to organise such capital (Murakami et al. 1979). Mechanisms to enable human capital accumulation and succession were advanced among the noble and samurai classes as well as among the peasantry by the end of sixteenth century (Teranishi 2014).

2.2 SUZUKI Shōsan's Vocational Ethics

The changes in, or the Japanisation of, Buddhism, described above, were reflected in the labour philosophy and vocational ethics advocated by SUZUKI Shōsan 鈴木正三 (1579–1655), who some consider to be the founder of Japanese vocational ethics (Shimada 1990), even though his teachings were not institutionalised into a movement or school (Ooms 1985). Shōsan was born into a samurai family. In the age of transition, between times of war and peace, he served with distinction as a warrior as well as a bureaucrat, and became a Zen Buddhist priest when he was forty-two years old. Even though he became a monk, he exhorted people to remain lay followers and to dedicate themselves to their family businesses as the best way to attain the truth of Buddha (Shimada 1990). On the other hand, Shōsan held the deep conviction that the Buddhist tradition could mediate a new social ethic, although he considered established Buddhism, in its contemporary form, useless (Ooms 1985).

Based on his awareness of human beings as *bonpu* and his understanding of the Mahayana idea of dependent co-origination (*engi* 縁起)[10] and indebtedness to or generosity of others (*on* 恩),[11] he emphasised the importance of rising above the self through submission to Buddha, and insisted that following the laws and principles of the fleeting world was the central practice of Buddhism. In other words, the world's teachings are the Buddha's teachings (*sehō soku buppō* 世法則仏法) and through the world's teachings one attains Buddhahood (Ooms 1985; Shimada 1990). The Avatamsa (*kegon* 華厳) Philosophy, a core thought of Mahayana Buddhism, inspired Shōsan to the understanding that every occupation was a manifestation of Buddha; that is to say, every occupation is an equally divine and valuable vocation (*tenshoku* 天職), and that occupational life led by *gudō*-ism is a bodhisattva path, or a means of achieving human perfectibility and spiritual salvation for the self as well as for others (Shimada 1990). Every member of society

[10] The idea of *engi* in Mahayana Buddhism means that all of the elements composing the world are insubstantial (emptiness or *kū* 空) while they arise mutually dependently (Bito 2000). In this respect, Buddhism is ontologically a system based in the category of *relatio* (Izutsu 2001: 23). A logical conclusion of this is that there is no need to discriminate something from something else.

[11] In the world of *engi*, all living things do not live independently but are allowed to live, thanks to other life as well as inanimate beings. A normative implication of this idea of indebted life is that human beings have a duty to live their life in gratitude for, and repayment of, their indebtedness to all other beings (*hō'on* 報恩). Ooms (1985) explained that Shōsan developed a decentralised definition of personality based on the idea of *on*; one is constructed by several centres outside of oneself.

living in the network of intricate mutual dependencies, or *on,* should comprehend that all occupations are Buddhist practice and that one ought to devote oneself to the occupation practiced by one's family (*kashoku* 家職), which provides an entrance into Buddhism, all the while knowing one's own station or *bun* 分 and not trespassing into others (Ooms 1985).

He preached to samurai on the importance of devoting their lives to selfless service to their lords on the basis of being aware of their functional responsibility for maintaining the social order and proactively developing their outstanding capabilities with both the sword and the pen. This way, they could carry out their duties in accordance with the highest ethical standards, centred on loyalty to their lords, as an expression of their gratitude for the favour conferred on them by their lords. As an extension of the ideal samurai ethic of courage and active service (Ooms 1985), Shōsan encouraged farmers to believe that those who engaged in agriculture assumed an innate divine mission to nurture and sustain all lives on earth and accumulated Buddhist merit through their heavy physical labour. Similarly, he taught craftsmen that they should devote themselves to their vocation, which intrinsically had a Buddha nature; craftsmen, on behalf of Buddha who vowed to realise the blissful happiness of the world, enabled people to go about their daily lives in comfort by providing them with necessary commodities. Shōsan recommended that merchants conduct their business, as a Buddhist practice, with integrity, diligence and frugality; commercial transactions aiming to make an honest profit while accounting for the best interests of their customers accorded with Buddha's mind of supreme compassion (*daihi* 大悲) (Shimada 1990).

3 Development of Vocational Ethics in the Tokugawa Era

3.1 Jugaku and the Vocational Ethics of the Samurai

After winning a decisive victory in the Battle of Sekigahara in 1600, Tokugawa Ieyasu 徳川家康 was appointed by Emperor Go-Yōzei 後陽成天皇 as the "shogun in charge of conquering barbarian territories". Based on this designation, Ieyasu set up his *bakufu* (shogunate government) in Edo (Tokyo) in 1603. This was the beginning of a 265-year-long period of peace after an era of war and treason that began with the Ōnin War in 1467. In 1615, the Tokugawa Shogunate established governance over the *daimyō* (feudal lords) in various parts of the country.

To strengthen its dominion, the Tokugawa *bakufu* developed both legal and political systems. The Acts for the Military Houses (the *Genna* Directives), aimed at controlling *daimyō* conduct, and the Acts Governing the Imperial Court and Nobility, which ensured the *bakufu*'s power to limit the behaviour of the imperial family and court nobles, were enacted in 1615. The policy of *sankin kōtai* obliged the *daimyō* to live in Edo and in their own domain, or *han*, for alternating years to show their loyalty to the shogunate. This policy was based on the Revised Acts for the Military Houses (the *Kan'ei* Directives), and institutionalised in 1635.

Christianity was banned in 1612 because it was considered to be a cause of domestic warfare. Moreover, it was believed that missionaries from the Catholic countries, particularly Spain and Portugal, had imperialistic territorial ambitions (Kanda 2005). *Terauke*, a kind of family registration system in which each member of a family had to be affiliated with a Buddhist temple, was implemented in 1635 to allow Japanese people to prove they were not Christians. For national security reasons, overseas voyages were subject to strict restrictions after 1633, and international trade was almost completely controlled by the *bakufu*. Among the Western nations, only Holland was permitted to maintain trade relations on the island of Dejima in Nagasaki, after the *bakufu* imposed an embargo on all trade with Spain in 1624 and Portugal in 1639. These defensive international policies were made possible through the abundant production and trade of gold, silver and copper, and by Japan's geopolitical position at the time. This contributed to an environment free from the threat of foreign invasion (Ōishi 1977).

The political regime of the Tokugawa *bakufu*, a military regime, was theoretically legitimised by Shintō, Buddhism and Confucianism,[12] reflecting Japan's tendency to use religious values as a means to achieve social and political ends (Ikegami 1995). These religions had amalgamated to a considerable degree, as demonstrated by an accompanying growing interchangeability of concepts, words and terms and by the spread of the theory of *honji suijaku* 本地垂迹 throughout Japan. This theory, which was formed at the end of the Heian Period, claimed that Shintō gods (*kami* 神) were manifestations of Buddhas and bodhisattvas (Inoue 1998).[13] According to Ooms (1985), Ieyasu's Testament (*Tōshōgū goyuikun* 東照宮御遺訓) used terminologies from Shintō, Buddhism and Confucianism. Revised versions of this document were widely read by samurai, at least until the end of the seventeenth century. The document states that the use of military force by the Tokugawa warrior rulers was a means of establishing a society in accordance with *tendō* 天道, or the Way of Heaven. The principle of *tendō* dictates that loyalty to superiors is the cardinal virtue required of retainers and selfish desire is a betrayal of *tendō,* meaning that samurai, as rulers, should display universal benevolence and generosity of heart towards the people.

Confucianism had been studied by Buddhist monks centred on Zen Buddhists since its arrival in Japan, as it had been an indispensable element of Zen Buddhist thought developed in China during the Song Dynasty (Suzuki 1959). The laicised Buddhist monk FUJIWARA Seika 藤原惺窩 (1561–1619) established *jugaku* 儒学, or the study (*gaku* 学) of Confucianism (*jukyō* 儒教), as an independent field based

[12] To demonstrate the legitimacy of the *bakufu*, Ieyasu was posthumously apotheosised and called *Tōshō Daigongen* 東照大権現 (Great Incarnation Shining over the East) and *shinkun* 神君 (divine ruler). As Ooms (1985) pointed out, the legitimatisation led to a social misrecognition of power and domination as virtue.

[13] Alternatively, the Yoshida Shintō sect insists that Buddhas are manifestations of *kami* (Bito 2000).

mainly on the Southern Song Neo-Confucian thought of ZHU Xi (*Shushi* 朱子 in Japanese), or the Cheng-Zhu school, who had formulated a new, far-reaching theory of Confucianism. This theory fused the logical systems of Buddhism and the Daoist notion of non/great-polar.

HAYASHI Razan 林羅山 (1583–1657), Seika's pupil, became a member of the strategic staff in the Tokugawa *bakufu* upon the recommendation of Seika, but he did not exert significant political influence during his lifetime (Ooms 1985). He interpreted ZHU Xi's theory as the essential way to legitimise the Tokugawa feudal system, which was structured by a class system that ranked samurai at the top, followed by farmers and townspeople composed of craftsmen and merchants.[14]

This class system was constructed on the basis of *ie* occupations, and human mobility between these classes was relatively high. Bito (2000, 2006) characterised the social structure of the Tokugawa feudal system as a "system of duties," which were assigned to the *ie*. The Japanese term *ie* is usually translated as a family or house(hold); however, *ie* in the Tokugawa Era was an organisation or a social unit, which had its own occupation (*kagyō* 家業 or *kashoku*) or duty (*yaku* 役) and property (*kasan* 家產), and was not necessarily categorised as a kin group but, rather, as a cooperative group centred on its *yaku*. Therefore, an employee was considered an *ie* member, and it was not unusual for an adopted son with no blood ties to the head of *ie* to take over the *kagyo* or *yaku*. According to Bito (2000), *yaku* involves the idea of great pride relating to *ie* members' duty of taking social responsibility based on their awareness of being a member of the *ie* and the society.

In the middle of the seventeenth century, when the *bakufu* saw a need to instil certain ethico-political values in the samurai (Ooms 1985), *jugaku* or *Shushigaku* 朱子学 became one of the most influential political doctrines of the Tokugawa *bakufu*. In peacetime, the political status of civilian personnel tended to rise, so *Shushigaku* was studied by a wider range of samurai who wanted to further their careers as bureaucrats[15] in the *bakufu* as well as in the local *han*. This resulted in the cultivation of competent *jugaku* scholars and policymakers such as ARAI Hakuseki 新井白石 (1657–1725) who was a close policy advisor to the sixth shogun TOKUGAWA Ienobu 徳川家宣, and in the dissemination of *jugaku* thought and ethics among the entire samurai class.[16]

A key feature of *Shushigaku* that developed in Tokugawa Japan was the importance of *chū* 忠, or loyalty to one's lord or superior, over *kō* 孝, or filial piety. Polity

[14] By 1871, there were outcaste people called *kawara no mono* (riverside residents) or *eta* (much polluted people), *hinin* (non-human) and *shōmoji*, who engaged in specific occupations such as leather manufacturing and performing arts and formed their own guilds (Hattori 2012).

[15] In the Tokugawa Era, the transformation of the samurai from a class of semiautonomous warriors into domesticated bureaucrats progressed while suppressing the samurai's resistance based on their honorific individualism (Ikegami 1995).

[16] Studies examining *Shushigaku* in the Tokuwaga Era criticised ZHU Xi's ideas and thought, leading to a regard for the Lu-Wang school of Neo-Confucianism as well as for original Confucianism. This also involved the foundation of Kokugaku 国学, the study of Japan's ancient culture, thought and spiritual world-view, before the arrival of Buddhism, Confucianism and Daoism to Japan, by an investigation of the early Japanese classics and the ideas of Koshintō or original Shintō.

overrode the family; in cases of a conflict of loyalty, the first duty was to one's lord rather than to one's family (Bellah 1985; Ikegami 1995). Professional matters were considered *ōyake* 公 (public) for samurai, whereas family matters were *watakushi* 私 (private).[17] Carrying out one's responsibility to one's superior was called *hōkō* 奉公, and meant that one dedicated oneself to *ōyake* matters without consideration of *watakushi* matters. *Hagakure*, the so-called "sacred text of *bushidō* 武士道," which was edited in 1716 and begins with the famous phrase, "The Way of the warrior (*bushido*) is found in dying" (Yamamoto 2014), sets out the knowledge and attitude required for a samurai to carry out his *hōkō* during peacetime, even though samurai were still forced to commit seppuku (ritual suicide) if they were involved in a quarrel[18] or a situation that stained another samurai's honour,[19] such as a house fire caused by negligence (Koike 1999).

Razan emphasised that the principles of the universe governed not only the order of nature but also the order of human society, including social strata. Human beings should seek to achieve an intuitive realisation of the principles of the universe through self-discipline. This idea, in conjunction with pre-existing aspects of samurai culture such as Zen Buddhism and the tea ceremony, provided samurai with the spiritual criteria to devote themselves to their occupations as prescribed by the hierarchical feudal system. In addition, the primacy of *chū* enabled the Tokugawa government to focus social action on the vertical ties of loyalty between superiors and inferiors (Bellah 1985). These ties led samurai to recognise *hōkō* as rational behaviour and encouraged them in their devotion to work and thrift. This relationship led to the development of a kind of vocational ethics for the samurai.

Bellah (1985) described how the obligation to fulfil one's status in the feudal hierarchy almost imperceptibly merged into the obligation to perform one's daily work. He also pointed out that the central importance of the polity and obligations to superiors during the Edo Period led to the prioritisation of *ōyake* interests over *watakushi* interests. This was normally expressed through the appropriate behaviour of each status group, or through knowing one's *bun*. Behind the awareness of one's *bun* was the Japanese religious belief that human beings live with the blessings (*on* 恩) of Supreme Beings, including the Heavens (*otentō-sama*), Buddha (*hotoke-sama*) and Shintō deities (*kami-sama*), as well as all other beings, and thus have a duty to make return for those blessings or *hō'on* 報恩 (Bellah 1985).

These vocational ethics were reflected in the samurais' faithfulness to their occupational duties. Although most samurai occupations in the hierarchical feudal system were hereditary, a samurai's occupation was not necessarily inherited by his

[17] On the concepts of *ōyake* and *watakushi*, see Nakada and Tamura (2005), who explain the nuanced differences between these Japanese concepts and their English counterparts.

[18] In the Tokugawa Era, the common law of *kenka ryōseibai* (all parties to a quarrel earn equally severe punishment) was observed (Ikegami 1995).

[19] The Tokugawa samurai culture of valuing honour above all – even above their lives – functioned as an ideology of the master-follower relationship by supplying an internal principle of predictability for judging the follower's and the master's actions and intentions. This ideology often became a tool in the hands of masters to encourage their followers' loyalty to them (Ikegami 1995).

biological son. Based on the system of vocational ethics and also on the Shintō-Confucian concept of filial piety to one's ancestors, when a father did not consider his biological son capable of achieving a satisfactory level of performance in his occupation, he could try to adopt a more competent person, who was not necessarily a sib, as his heir.

Those holding high status or important positions were required to fulfil moral vocational obligations as well. *Daimyō* had a responsibility to their people as well as to the *bakufu* and the shogun. When a lord did not appropriately perform his tasks, his retainers, especially the *karō* (chief retainer), could expostulate with him. It was risky for a retainer to take this action in the stratified feudal society, for it occasionally resulted in their seppuku or *oteuchi* (decapitation). However, such expostulation was considered a vocational duty of samurai retainers, and was a *karō*'s right. If the expostulation did not improve the lord's performance, his retainers could decide to commit *shukun oshikome*, a practice in which the retainers placed their lord under house arrest, deposed him and replaced him with an appropriate relative (Kasaya 2006). This practice exemplified the subordination of *watakushi* interests to *ōyake* interests in the samurai psyche. In a samurai's life, duty and family welfare were valued less than the order and welfare of the *han* society.

Bellah (1985) asserted that particularism held primacy in Tokugawa Japan. Commitment to a particular system or collective, whether it was family, *han,* or Japan as a whole, took precedence over universal values such as truth and justice. The importance of the collective and of one's particular relation to it was demonstrated by the enormous symbolic importance of the head of the collective, whether it be the head of the family, the feudal lord, or the emperor. One's particular allegiance to the head was as a symbol rather than as a person. This kind of relationship implies the possibility of holding deep loyalty to a person, such as the emperor or shogun, without having any personal connection at all. Bellah (1985) posited that this generalised particularism was, in some ways, the Japanese functional equivalent to the universal ethic of Protestant Christianity that, as Weber (2003) suggested, contributed to the modernisation of Western states. Particularism implies that the primary concern is the overarching goal of a particular system or collective. Hence, loyalty in Japan meant not passive devotion alone, but active service and performance.

3.2 Zen Buddhism as the Basis of Samurai Morality

Considering the long tradition of Japanese Zen Buddhism from the end of the twelfth century, it is not surprising that *Shushigaku* was smoothly accepted by samurai in the Tokugawa Era as the foundation for their ethical thinking and behaviour. Confucianism had been studied by Zen Buddhist monks, and many Confucian ideas, especially those concerning practical ethics, had been assimilated during the independent development of Zen Buddhism in Japan.

The development of Zen Buddhism during the Sui and Tang Dynasties in China was significantly influenced by the Daoist thought of Laozi (*Rōshi* 老子 in Japanese) and Zhuangzi (*Sōshi* 荘子 in Japanese). These Daoist gurus claimed that everything in the universe is unified, that language is inadequate for perceiving or telling the truth, and that experiential intuition alone allows people to understand the *Dao* 道 (the Way). They contended that artificiality is an obstacle to knowing the world as it is (Mori 1994). These Daoist ideas are reflected in Zen training, which consists of personally experiencing the truth rather than appealing to intellectualism or to systematic theorisation, and a position illustrated by the Zen motto of "no reliance on words" or *furyū monji* 不立文字 (Suzuki 2007).

Samurai began to adopt Zen Buddhism as their moral and philosophical basis in the Kamakura Era, during which Japan's first *bakufu* governed the country; it was also during this era that the Mongols invaded (in 1274 and in 1281). SUZUKI Daisetsu 鈴木大拙, a Zen Buddhist philosopher, explained why samurai in that period had faith in Zen Buddhism:

> Zen has sustained them [samurai] in two ways, morally and philosophically. Morally, because Zen is a religion which teaches us not to look backward once the course is decided upon; philosophically, because it treats life and death indifferently. This not turning backward ultimately comes from the philosophical conviction; but, being a religion of the will, Zen appeals to the samurai spirit morally rather than philosophically. From the philosophical point of view, Zen upholds intuition against intellection, for intuition is the more direct way of reaching the Truth. Therefore, morally and philosophically, there is in Zen a great deal of attraction for the military classes...
>
> Secondly, Zen discipline is simple, direct, self-reliant, self-denying; its ascetic tendency goes well with the fighting spirit... A good fighter is generally an ascetic or stoic, which means he has an iron will. This, when needed, Zen can supply. (Suzuki 1959: 61–62)
>
> ... The notion of death, on the one hand, makes one's thought extend beyond the limitations of this finite life, and, on the other hand, screws it up so as to take daily life seriously. It was, therefore, natural for every sober-minded samurai to approach Zen with the idea of mastering death. Zen's claim to handle this problem without appealing either to learning or to moral training or to ritualism must have been a great attraction to the comparatively unsophisticated mind of the samurai. There was a kind of logical relationship between his psychological outlook and the direct practical teaching of Zen. (Suzuki 1959: 72)

Zen Buddhism is an intellectual religion, but it denies the use of language for logical analysis to accomplish the aim of Zen, which is perceiving the world as it is, as the central fact of life. Daisetsu elaborates on this purpose:

> The great truth of Zen is possessed by everybody. Look into your own being and seek it not through others. Your own mind is above all forms; it is free and quiet and sufficient; it eternally stamps itself in your six senses and four elements. In its light all is absorbed. Hush the dualism of subject and object, forget both, transcend the intellect, sever yourself from the understanding, and directly penetrate deep into the identity of the Buddha-mind; outside of this there are no realities. (Suzuki 1964: 16)

During the turbulence of the Kamakura Era, samurai were repeatedly forced to face crises through warfare. Zen Buddhism provided them with a direct and intuitive way of addressing difficult existential problems. Through mental and physical practices such as *zazen*, they were encouraged to experience *satori* 悟り/覚り

(enlightenment) without any external aid or interference. This individual religious experience lies at the core of Zen Buddhism.

In the state of *satori* one maintains *mushin* 無心 (emptiness of mind and deference to the absolute being)[20] and thereby attains *anjin* 安心 (peace of mind) (Suzuki 2007). The *satori* experience enables the practitioner to realise the oneness of human beings and the universe, as well as the oneness of life and death; thus, the practitioner transcends his death and rids himself of his ego. *Satori* awakens one to the truth that everything in the universe is unified and indistinguishable. Thus, the person of *satori* has overcome and transcended the limitations of an arbitrarily articulated world and has attained the stage of absolute freedom in articulating the non-articulated (Izutsu 2001). Once a samurai experienced the state of *satori*, he could fight without hesitation or anxiety.

During the peaceful Tokugawa Era, most samurai continued to practice Zen Buddhism, even though *jugaku* scholars criticised Buddhism and drove it from the government (Suzuki 2007; Ooms 1985). The idea of the oneness of human beings and the universe led to the conclusion of Mencius, a Confucian scholar of Chinese antiquity: that human nature was fundamentally good. Because *satori* was the ideal state of mind for the samurai, *mushi* 無私 (no-ego, and therefore, no-*watakushi*) was recognised as the supreme virtue. It is evident that *mushi* was the basis of the vocational ethics and particularism described above. Here, *mushi* does not mean denial or annihilation of the self, but rather, affirmation of the self or discovery of one's true self, that is, when the self is not the self, the self is the true self.[21] An intuitive understanding of this seemingly illogical Zen proposition, the logic of "non-therefore" or "identity of contradiction" (Nakamura 2009),[22] through religious experience leads to the realisation of the oneness of the self and of others; therefore, a single human being is inseparable from others and from society (*ōyake*), and an altruistic act is thereby a selfish act.

The philosophy of Zen Buddhism affected Japanese ways of thinking for a long time; philosophers in the modern age are no exception. For example, Nishida Kitarō 西田幾多郎 (1870–1945), one of Japan's most prominent modern philosophers, contended, based on his Zen practice, that the self can attain correct knowledge of the Being of beings when it transforms from the everyday self grounded in *ba* (place) of being into the self as *ba* based on the *ba* of nothing (Yuasa 1987). Watsuji Tetsurō 和辻哲郎 (1889–1960) stated that the Japanese term *ningen*, which corresponds to the English concept of a human being, is used to express the *seken* (world) a person lives in and, simultaneously, a person who lives in a *seken*. People become human beings when they reside in a *seken*, and are recognised as

[20] Izutsu (2001) pointed out that *mushin* is not to be understood in a purely negative sense as the mind in the state of torpidity and inertness or sheer ecstasy; it is a psychological state in which the mind finds itself at the highest point of tension, a state in which the mind works with utmost intensity and lucidity to the contrary, despite its apparent inaction.

[21] According to Izutsu (2001), the Oriental "Nothingness (*mu* 無)" is not a purely negative ontological state of there being nothing, but it is a plenitude of Being.

[22] The general form of this logic is "A is not A, therefore it is A" or "A is A, because A is non-A."

human beings because they represent the whole *seken*. Watsuji maintained that the principal problem of ethics is the fundamental dualistic structure of *ningen*. According to Watsuji, *ningen* is the self plus *aidagara* (relationships with others), and the inseparability of human existence from *aidagara* must be understood (Watsuji 2007).

3.3 Vocational Ethics of the Merchant Class

The ethical values best typified by the samurai class, often referred to as *bushido*, became quite general and were institutionalised among all classes by the end of the Tokugawa period. For example, *mushi* service of the collective goal was characteristic of the commercial houses. The *ōbantō* (head clerk) of a commercial house, who corresponded to the *karō* of a *han*, had the greatest responsibility for the business of the commercial house. If the master were dissolute, it was the *ōbantō*'s duty to restrict him and maintain the standing of the business. When necessary, an *ōbantō* did not hesitate to sequester his master. Performance values were clearly predominant over status values (Bellah 1985). As in the samurai class, the biological son of the master of a commercial house did not necessarily inherit the family business. When the master, and his *ōbantō*, considered the biological son incapable of a satisfactory level of performance in the business, a more competent man could be adopted into the family.

The development of trade and manufacturing in the Tokugawa Era as well as increased demand for documentation management among farmers and townspeople led to the widespread establishment of small private schools called *terakoya* or *tenarai-shinanjo* for non-samurai children, while the children of samurai families studied the cultural, moral and martial subjects necessary for their duties at *han* schools. Takahashi (2007) estimated that the number of *terakoya* was more than sixty thousand, and that the literacy rate in Japan was more than seventy per cent by the end of the Tokugawa Era. *Terakoya* masters taught their pupils proper decorum, reading, writing and abacus arithmetic using a variety of texts including the *Thousand Character Classic* and the Five Classics and the Four Books of Confucianism. Many *shijuku* (private schools) that provided a higher level of education for both samurai and commoners were established by *jugaku* scholars like NAKAE Tōju 中江藤樹 (1608–1648), who studied *Yōmeigaku* 陽明学 based on the Ming Neo-Confucian philosophy of WANG Yangming (*Ō Yōmei* 王陽明 in Japanese) and was consulted by *daimyō* on policymaking, and OGYŪ Sorai 荻生徂徠 (1666–1728) who criticised *Shushigaku* but became a policy advisor of the *bakufu* anyway. Kaitokudō, the Confucian merchant academy where education based on the belief that morality and economy were inseparable was provided (Najita 2009), was set up in 1724 by five business tycoons in Osaka. The relatively low status of the Confucian scholars in the Tokugawa political pyramid provided them with intellectual flexibility. At the same time, that status enabled the Confucian scholars to spread Confucian moral philosophies among non-samurai classes and encouraged the development of non-samurai Confucian scholars (Ikegami 1995).

A theory of vocational ethics for the merchant class in the Tokugawa Era was established by ISHIDA Baigan 石田梅岩 (1685–1744) and his disciples including TEJIMA Toan 手島堵庵 (1718–1786). Their doctrine of merchant ethics or *shōnindō* 商人道 (the way of merchants) is called *Sekimon Shingaku* 石門心学 (the Ishida School of the study of the mind). Baigan was born into a farming family in a small village near Kyoto and studied Shintō, Buddhism and Confucianism while working for his family business and a local draper. Following his retirement, he achieved enlightenment through Zen practice at the age of forty-four, and set out to edify the merchant class. His thoughts were based primarily on ZHU Xi's interpretations of the Mencius concepts of heart and nature (Bellah 1985; Shimada 1990; Yui 2007).

In his seminal book *Tohi Mondō* (dialogues between an urban or learned man and a countryman or an unlearned man), Baigan prescribed how merchants ought to conduct themselves. Merchants who do not understand *shōnindō* seek excessive profits, resulting in the bankruptcy of their family business. Those who understand *shōnindō* never become greedy, devote themselves to their business with benevolence and, consequently, enjoy prosperity. Merchants should treat their customers in a cordial manner, just as samurai do not hesitate to sacrifice their personal lives for their lord. Amassing a fortune through saving pennies and serving customers in good faith while maintaining the virtue of thrift is truly consistent with *shōnindō*, which has the core values of fairness and honesty in commercial trade. Honesty for merchants means the open gain of a fair profit. The basic morality required to manage a commercial house is a master's benevolence to his employees, and their reciprocal loyalty to the master. Harmony among the master and his employees, as well as self-discipline, is necessary for the conduct of moral commercial endeavours (Bellah 1985; Yui 2007).

The philosophy of *Sekimon Shingaku* was accepted by merchants, especially those who came from the samurai class, as a theory to legitimise commercial activities. Baigan described this idea clearly:

> Obtaining profit from sale is the Way of the merchant. I have not heard selling at cost called the Way... The merchant's profit from sale is like the samurai's stipend. No profit from sale would be like the samurai serving without a stipend. (translated into English by Bellah (1985: 161))

This thought was further supported by a wide range of people beyond the merchant class in conjunction with the *hōtoku* movement in rural areas (see the next subsection), given that *Sekimon Shingaku* was based on the Japanese attitude that integrated Shintō, Buddhism and Confucianism and valued the key virtues of sincerity and task performance (Yui 2007). Vocational ethics for craftsmen, who had developed their ethos to put more weight on earning their place in history through their work than on earning money, and for farmers were also influenced by *Sekimon Shingaku*; they were encouraged to know their *bun* and dedicate themselves to their vocational duties, which were decreed by heaven (Bellah 1985; Shimada 1990).

3.4 Vocational Ethics of Farmers

NINOMIYA Sontoku 二宮尊徳 (1787–1856), who was from the farmer class, was recognised as a great agricultural policymaker and practitioner due to his remarkable achievements in agrarian rehabilitation. Sontoku lived in the late Tokugawa Period during which Japanese rural populations were ravaged by famines caused by natural disasters, political turmoil and a heavy tax burden. Abandonment of cultivation, abandonment of farming villages and moral degeneration among farmers was not unusual throughout Japan at the time. Peasant uprisings were often staged (Shimada 1990). Many types of *kō* 講, or cooperatives, were organised as self-help initiatives from below to help farming villages survive famines (Najita 2009).

Sontoku founded the *hōtoku* 報徳 (return for virtue)[23] movement with the support of the Odawara-*han* to rebuild devastated rural areas based on his conviction that raising the level of morality of farmers and improving their economic productivity was his duty (Bellah 1985). Najita (2009) pointed out that the central meaning of *hōtoku* is "work as virtue."

Sontoku's idea of *hōtoku* was developed from his independent study of Shintō, Buddhism and Confucianism as well as his agricultural experience. It dictated that a human being must be diligent in exchange for the virtue of heaven, earth and people, and that maintaining one's humanity (*jindō* 人道; the way of human beings) through diligent labour, with integrity (*shisei* 至誠) and altruism (*jin* 仁), made it possible for one to enjoy true affluence. According to Najita (2009), as opposed to the moral philosophy of *Shusigaku*, Sontoku held that nature was not the model of *jindō*, rather, heaven or universal nature resided in each distinct self as the energy to realise truthfulness and virtue. For Sontoku, agriculture, a set of activities to nourish life, was a moral responsibility and an imperative for human beings; thus, *jindō* and virtue were realised through the quality of agricultural work within a cooperative or contractual framework.

Hōtoku shihō 報徳仕法, or agrarian rehabilitation plans that Sontoku devised, were centred on the principles of *bundo* 分度, a rational economic plan for rehabilitation, and *suijō* 推譲, utilising surplus for households' savings, which could be used to meet unexpected expenses as well as social capital building and improvements in local communities. Proper implementation of *bundo*, in which farming people were required to lead their lives within their *bun*, was a prerequisite for creating surplus and, thus, for practicing *suijo*. He stipulated that the inhabitants of the rural areas he was asked to rebuild had to maintain *jindō*. In particular, they were required to behave sincerely, take care of themselves, know their own *bun*, practice economy, and offer their agricultural surpluses to improve the living conditions of them and others in the area (Shimada 1990).

[23] Bellah (1985) explained that *hōtoku* means essentially the same thing as *hō'on*.

4 Restoration and Revision of Japanese Core Vocational Ethical Values

Following the exploration of the core vocational ethics developed in the Tokugawa Era, let us consider why it is necessary to, and how we can, restore and revise them in today's ICT-dependent society. As Weber (2003) predicted, a lot of nullities – specialists without spirit and sensualists without heart – have recently emerged in Japan, where capitalist economic development has succeeded to a remarkable degree. The idea of diligently earning an honest penny and practicing economy was often flouted during the economic bubble from the mid-1980s to the early 1990s. After the bubble burst, arguments based on neoliberalism or market fundamentalism have continued to influence the media and public opinion. It seems that greed, underpinned by market fundamentalism, has already superseded market ethics. Moreover, in today's capitalist society, market ethics seem to govern civil ethics, and civil ethics seem to govern the fundamental ethical values that prescribe individual judgment between good and evil and right and wrong. For example, many businesses consider business ethics and corporate social responsibility as instruments to ensure sustainable profitability (Murata 2010). The economic attractiveness of China's huge consumer market has overshadowed previous concerns over domestic corruption and the suppression of human rights. The ideas of efficiency-first and commercialism have prevailed in the Japanese society.

It is natural that core ethics govern civil ethics and civil ethics govern market ethics. Unfortunately, because the Japanese have lost sight of their core vocational ethics, the reverse is now true. This reversal has made it difficult for Japanese people to address ethical issues concerning ICT and ICT-based information systems, most of which have been developed and operated by business organisations, because addressing such issues tends to be recognised as a cost factor or a waste of time and money. The revival of core vocational ethics and a reversal of the aforementioned situation are necessary to appropriately cope with ethical problems associated with technology.

During the Meiji Restoration, democracy, constitutional politics and industrialisation were achieved through the repeated affirmation – and, when necessary, the revision – of the Japanese core vocational ethics developed during the Tokugawa Period. For example, Shibusawa Eiichi 渋沢栄一 (1840–1931), one of the preeminent businessmen of modern Japan, advocated undertaking business based on Confucian thought as described in the Analects, and criticised the status-quo-oriented aspects of *shōnindō* (Shibusawa 2008). Fukuzawa Yukichi 福澤諭吉 (1835–1901), a great enlightenment thinker and educator, stressed the necessity and importance of the general public's acquiring knowledge from Western studies, based on his deep understanding of Japanese core vocational ethics (Fukuzawa 2012). However, these attitudes became rare following the decline of democracy, the subsequent rise of the military, and the penetration of totalitarianism in the early Showa Era.

After the Second World War, the process of restoring democracy and constitutional government in Japan did not address Japanese core vocational ethics. Primary and secondary educational programmes have made little mention of these values, partly because they were considered to be among the causes of Japanese totalitarianism. As a result, the Japanese people now face a superficial, rootless democracy, and have lost sight of their core vocational ethical values. This, in addition to the tendency of the Japanese to consider any discussion of ethics as a matter for children, has led to a situation in which Japanese people have almost lost the capacity for ethical discourse in their vocational lives. The results of the surveys of Japanese youngsters' attitudes to work conducted by the Japan Productivity Center in 2016 illustrate that situation. Among 1,280 respondents, only 9.3% stated that their object of labour was generating benefits for society, whereas 41.7% answered they worked for the purpose of leading an enjoyable life and 27.0% said they worked to become economically affluent showing their indifference to the connection between their jobs and the society (Japan Productivity Center 2016b). To the question "Do you carry out an ethically questionable instruction of your superior which would provide your company with a benefit against your conscience?", 45.2% of 1,951 respondents responded "Yes", 10.6% expressed their opposition saying "No", and the remaining 44.2% responded "I have no idea" (Japan Productivity Center 2016a). Those who save their own skin and adopt a particularistic attitude in their vocational lives can never be professionals. Professionals are required to recognise that their work is primarily a form of public service and that public interest should guide their judgment and decision-making, and to accept full responsibility for their work and remain honest with themselves and with others (Murata 2013).

Consequently, the Japanese are prone to think of ICT as just a technological matter that has no relationship with social and ethical issues. Even when they are aware of ethical issues regarding ICT, the Japanese address the issue by seeking out a similar case and apply the same solution adopted in that situation. This indicates that they have a lack of ability to proactively cope with novel issues caused by the development and use of ICT. Moreover, almost all of the educational programmes for engineering ethics provided at Japanese engineering colleges are developed based on those used for American students, without any consideration of traditional Japanese vocational ethics. Such programmes are ineffective at creating engineers who can appropriately address ethical issues caused by the development and usage of technology as professionals, who have to take total responsibility for their work and put the public interest first as described above. Furthermore, they are not adequately prepared to simultaneously consider the local nature of humans in relation to the global natures of technology itself. Japanese ICT engineers, as professionals, are required to find renewed meaning of, for example, *ōyake*, *hōkō* and *mushi* suitable to the current ICT-driven globalised world based on a correct understanding of the traditional notions of them.

It is urgent that the Japanese people restore their core vocational ethical values, but it is also necessary to understand the limitations of these ethics and to implement necessary revisions. Traditional vocational ethics in Japan encouraged people to know their *bun* and to focus only on their own duty, without broadening their perspective. This emphasis may have led to an obedient attitude towards authority, as observed in *okami-ishiki* (the Japanese tendency to blindly trust the government), and the status quo. In addition, the extent of *ōyake* can be set too narrowly. In fact, Japanese particularism relates to the concept of *seken*. Abe (1995) pointed out that an ordinary Japanese person lives, at least mentally, not in society but in *seken*, a small world defined by relationships of duty and empathy. Ikegami (1995) showed the connection between the modern Japanese sentiment of honour, with one's social appearance associated with one's status in the eyes of the *seken*, and the ethic focused on avoiding ridicule in their *seken* among the samurai in the Kamakura Period. The vertical orientation of Japanese society (Nakane 1970) and Japanese insular collectivism (Adams et al. 2009) can be understood in this context. Moreover, the recent changes of the Japanese economic environment, such as the collapse of the lifetime employment system, have led to a deterioration of the function of the modernised *seken* centred on companies as modernised *ie* and, thus, institutions of social norms (Murata and Orito 2016).

Nonetheless, considering the fact that a majority of ICT engineers work for business organisations, traditional Japanese virtues such as *mushi* and *hō'on* and ideas like *engi* are useful as a basis for addressing ethical issues concerning ICT. Ethical consideration and discourse based on these virtues and ideas can be accepted relatively easily by contemporary Japanese because they have been incorporated into Japanese culture. Actually, Japanese business people tend to consider private companies as public institutions and think that they work for their companies and conduct their business thanks to *engi*. Such virtues and ideas would not conflict with those associated with other religious cultures such as Christianity and Islam. The global nature of ICT implies a multicultural character for ethical issues concerning technology. Thus, for Japanese people, a deep understanding of their core vocational ethical values and of the limitations of these values is essential for addressing such issues. Those who take part in an ethical discourse in a multicultural context are required to appreciate their own ethical foundations and relativize them. If the participants fail to accomplish this, the discourse may cause mutual misunderstanding and conflict among the participants.

The educational programmes for ICT engineers to learn their own vocational ethical traditions and relativize them have to be set up and conducted not only at engineering colleges but also at companies as part of their in-house training programmes, given the aforementioned attitudes of Japanese young employees towards work and the fact that ICT engineers did not necessarily study at engineering colleges. An appropriately professional working environment for ICT engineers, where decisions made by them as professionals are respected, has to be constructed based on the careful examination and necessary revisions of Japanese traditional vocational ethics to ensure the autonomy of ICT engineers and to make their professional judgments into a reality.

5 Conclusions

In the *Analects* of Confucius, it is written that, "If a man keeps cherishing his old knowledge, so as continually to be acquiring new, he may be a teacher of others" (Wei Zheng 2: 11) (Legge 1971). ICT engineers and policymakers who wish to fulfil their professional responsibilities should consider the contemporary implications of this statement in the context of the development and use of ICT and ICT-based information systems. The global nature of ICT necessitates a multicultural approach to the ethical issues associated with technology. To cope with these issues in a multicultural context, the Japanese people must recognise their own core vocational ethical values in order to relativize these values and conduct ethical discourse. Japan is a global leader in ICT, but the Japanese have lost sight of their core vocational ethical values. The restoration of Japanese core vocational ethical values as a basis for ethical thinking and discourse, as well as the recognition of these values' limitations, constitutes the first steps towards appropriately addressing ethical issues concerning ICT.

Acknowledgements This study was supported by the MEXT (Ministry of Education, Culture, Sports, Science and Technology, Japan) Programme for Strategic Research Bases at Private Universities (2012–2016) project "Organisational Information Ethics" S1291006 and the JSPS Grant-in-Aids for Scientific Research (B) 25285124.

References

Abe, Kinya 阿部謹也. 1995. *What Is "Seken"?* 「世間」とは何か. Tokyo 東京: Kōdansha 講談社.

Adams, Andrew A., Kiyoshi Murata, and Yohko Orito. 2009. "The Japanese Sense of Information Privacy." *AI and Society* 24.4: 327-341.

Analects. 1971. In *Confucius: Confucian Analects, the Great Learning and the Doctrine of the Mean*, trans. by James Legge. New York: Dover.

Bellah, Robert N. 1985. *Tokugawa Religion: The Roots of Modern Japan*. New York: Free Press.

Bito, Masahide 尾藤正英. 2000. *The History of Japanese Culture* 日本文化の歴史. Tokyo 東京: Iwanami Shoten 岩波書店.

Bito, Masahide 尾藤正英. 2006. *What Is the Edo Era? The Early and Late Modern Period in Japan* 江戸時代とは何か-日本史上の近世と近代. Tokyo 東京: Iwanami Shoten 岩波書店.

Fukuzawa, Yukichi. 2012. *An Encouragement of Learning*, trans. by David A. Dilworth. Tokyo: Keio University Press.

Hattori, Hideo 服部英雄. 2012. *Kawara No Mono, Hinin, Hideyoshi* 河原ノ者・非人・秀吉. Tokyo 東京: Yamakawa Shuppansha 山川出版社.

Ikegami, Eiko. 1995. *The Taming of the Samurai: Honorific Individualism and the Making of Modern Japan*. Cambridge, MA: Harvard University Press.

Inoue, Nobutaka 井上順孝 ed., 1998. *Shintō: A Japanese Religious System* 神道 日本生まれの宗教システム. Tokyo 東京: Shinyōsha 新曜社.

Izutsu, Toshihiko. 2001. *Toward a Philosophy of Zen Buddhism*. Boston, MA: Shambhala Publications.

Japan Productivity Center 日本生産性センター, 2016a. A New Employee Attitude Survey in Spring 2016 2016年度 新入社員 春の意識調査. Available at http://activity.jpc-net.jp/detail/mdd/activity001476/attached.pdf (Accessed 5 May 2017).

Japan Productivity Center 日本生産性センター, 2016b. A Survey of New Employee Awareness of Work in 2016 平成28年度 新入社員「働くことの意識」調査結果. Available at http://activity.jpc-net.jp/detail/lrw/activity001478/attached.pdf (Accessed 5 May 2017).

Kanda, Chisato 神田千里. 2005. *The Shimabara War: The Faith in Christianity and Armed Uprising* 島原の乱 キリシタン信仰と武装蜂起. Tokyo 東京: Chūōkōron-Shinsha 中央公論新社.

Kasaya, Kazuhiko 笠谷和比古. 2006. *The Structure of Shukun Oshikome: Daimyō and Their Retainers* 主君「押込」の構造 近世大名と家臣団. Tokyo 東京: Kōdansha 講談社.

Koike, Yoshiaki 小池喜明. 1999. *Hagakure: Bushi and "Hōkō"* 葉隠-武士と「奉公」. Tokyo 東京: Kōdansha 講談社.

Mori, Mikisaburō 森三樹三郎. 1994. *Laozi, Zhuangzi* 老子・荘子, Tokyo 東京: Kōdansha 講談社.

Murakami, Yasusuke 村上泰亮, Kumon Shumpei 公文俊平, and Satō Seizaburō 佐藤誠三郎. 1979. *The Ie Society as Civilisation* 文明としてのイエ社会. Tokyo 東京: Chūōkōron Sha 中央公論社.

Murata, Kiyoshi. 2010. "Information Ethics in Business Organisations: Socially Responsible Development and Use of Information and Communication Technology." *Journal of Information and Management* 30.3: 48-57.

Murata, Kiyoshi. 2013. "Construction of an Appropriately Professional Working Environment for IT Professionals: A Key Element of Quality IT-Enabled Services." In *IT Enabled Services*, edited by Shiro Uesugi. Wien: Springer, 61–75.

Murata, Kiyoshi, Yasunori Fukuta, Yohko Orito, and Andrew A. Adams. 2017. "Few Youngsters Would Follow Snowden's Lead in Japan." *Journal of Information, Communication and Ethics in Society* 15.3: 197-212.

Murata, Kiyoshi, and Yohko Orito. 2008. "Rethinking the Concept of the Right to Information Privacy: A Japanese Perspective." *Journal of Information, Communication and Ethics in Society* 6.3: 233-245.

Murata, Kiyoshi, and Yohko Orito. 2016. "Communication Ethics in Japan: A Socio-cultural Perspective on Privacy in the Networked World." In *Ethics and Communication: Global Perspectives*, edited by Göran Collste. Lanham, MD: Rowman & Littlefield International, 163–180.

Najita, Tetsuo. 2009. *Ordinary Economies in Japan: A Historical Perspective, 1750–1950.* Berkeley/Los Angeles, CA: University of California Press.

Nakada, Makoto, and Takanori Tamura. 2005. "Japanese Conceptions of Privacy: An Intercultural Perspective." *Ethics and Information Technology* 7.1: 27-36.

Nakamura, Kojiro. 2009. "The Significance of Toshihiko Izutsu's Legacy for Comparative Religion." *Intellectual Discourse* 17.2: 147-158.

Nakane, Chie. 1970. *Japanese Society*. Berkeley, CA: University of California Press.

Ōishi, Shinzaburō 大石慎三郎. 1977. *The Edo Era* 江戸時代. Tokyo 東京: Chūōkōron-Shinsha 中央公論新社.

Ooms, Herman. 1985. *Tokugawa Ideology: Early Constructs, 1570–1680.* Princeton, NJ: Princeton University Press.

Orito, Yohko, and Kiyoshi Murata. 2008. Socio-cultural Analysis of Personal Information Leakage in Japan. *Journal of Information, Communication and Ethics in Society* 6.2: 161-171.

Orito, Yohko, Kiyoshi Murata, Yasunori Fukuta, Steve McRobb, and Andrew A. Adams. 2008. "Online Privacy and Culture: Evidence from Japan." Paper presented at ETHICOMP 2008, September 24–26, at the University of Pavia (*Proceedings of ETHICOMP 2008*: 615–622).

Shibusawa, Eiichi 渋沢栄一. 2008. *Analects of Confucius and Abacus* 論語と算盤. Tokyo 東京: Kadokawa Gakugeishuppan 角川学芸出版.

Shimada, Akiko 島田燁子. 1990. *Japanese Vocational Ethics* 日本人の職業倫理. Tokyo 東京: Yūhikaku 有斐閣.
Suzuki, Daisetsu T. 1959. *Zen and Japanese Culture*. Princeton, NJ: Princeton University Press.
Suzuki, Daisetsu T. 1964. *An Introduction to Zen Buddhism*. New York: Grove Press.
Suzuki, Daisetsu 鈴木大拙, 2007. *Mushin* 無心ということ. Tokyo 東京: Kadokawa Gakugeishuppan 角川学芸出版.
Takahashi, Satoshi 高橋敏. 2007. *Educational Ability in the Edo Period* 江戸の教育力. Tokyo 東京: Chikumashobō 筑摩書房.
Teranishi, Jūrō 寺西重郎. 2014. *Religious Foundations of Economic Behaviour in Japan* 経済行動と宗教 日本経済システムの誕生. Tokyo 東京: Keisōshobō 勁草書房.
Watsuji, Tetsurō 和辻哲郎. 2007. *Ethics as the Study of Man* 人間の学としての倫理学. Tokyo 東京: Iwanami Shoten 岩波書店.
Weber, Max. 2003. *The Protestant Ethic and the Spirit of Capitalism*, trans. by Talcott Parsons. New York: Dover.
Yamamoto, Tsunetomo. 2014. *Hagakure: The Secret Wisdom of the Samurai*, trans. by Alexander Bennett. Tokyo: Tuttle Publishing.
Yuasa, Yasuo. 1987. *The Body: Toward an Eastern Mind-body Theory*. Albany, NY: State University of New York Press.
Yui, Tsunehiko 由井常彦. 2007. *Tohi Mondō: The Way and Mind of Management* 都鄙問答 経営の道と心. Tokyo 東京: Nikkei Publishing 日本経済新聞出版社.

Chapter 8
Robots Seen from the Perspectives of Japanese Culture, Philosophy, Ethics and *Aida* (betweenness)

NAKADA Makoto

1 Introduction

It is often said that the ethical discussion on robots, robotics and HRI (human-robot-interaction) is poor in Japan. However, this is only a superficial assessment. Although topics such as "autonomy (of robots)' or "responsibility (of robots)" are not "hot" topics, Japan is a country where different views on robots are commonly accepted. These views are often based on cultural and social traditions. As the cases of "Astroboy (*Tetsuwan Atomu*)" or *Aibo* show, Japanese robots are usually thought to be friendly and supportive of humans. In Japan humans and robots are believed to share some important values such as "*mononoaware* もののあわれ" (the sensitive beauty, poetic beauty or the pathos of things and sensitivity to things) and even meanings of life and death. These views on robots reflect Japanese ways of understanding this world in which different entities such as humans and other beings are not separated from each other. Rather, these entities are in a state of primordial oneness. As authors and philosophers such as NISHIDA Kitarō 西田幾多郎, WATSUJI Tetsurō 和辻哲郎, YUASA Yasuo 湯浅泰雄 and ICHIKAWA Hiroshi 市川浩 suggest, various differences between "mind and body" or "subject and object" are thought to disappear in this authentic and primordial state of existence. Robots are regarded as entities that belong to this oneness when we see them from the "betweenness,"

The original version of this chapter was revised: MS Mincho font was updated throughout the book for Japanese and Chinese characters. The correction to this book is available at https://doi.org/10.1007/978-3-319-59027-1_13

NAKADA Makoto (✉)
Graduate School of Humanities and Social Sciences, University of Tsukuba, Tsukuba, Ibaraki, Japan
e-mail: nakadamakoto@msd.biglobe.ne.jp

Thomas Taro LENNERFORS, MURATA Kiyoshi (eds.), *Tetsugaku Companion to Japanese Ethics and Technology*, Tetsugaku Companions to Japanese Philosophy 1,
https://doi.org/10.1007/978-3-319-59027-1_8

"*aida* 間[1]" or "*rinri* 倫理[2]" perspectives suggested by these authors. In this sense, Japanese people and robots exist in a state of reciprocity or transaction which is itself an ethical matter, if we understand "ethics" as reflecting our existence grounded in "betweenness" or "*aida*."

In addition, Japanese attitudes toward robots seem to reflect Japan's hope for the robot industry as sources of economic and social development in the era after the "oil shocks" of the 1970s. Japanese positive attitudes toward industrial robots and humanoids seem to be related to these historical and social conditions.

This chapter will discuss people's ethical attitudes toward robots by examining the cultural-ethical perspectives behind these attitudes and the historical social conditions related to the robot industry after the 1970s. This analysis frames roboethics as something constructed by historical and social conditions, values and philosophical perspectives as well as technological circumstances. If viewed from the perspectives of "*aida*" or "betweenness," robots seem to represent aspects different from the "Western" dualism of subject and object. In addition, when robots evolve as a result of technological advancement or there is a shift in public interest, for example from industrial robots to social robots, this might change the ways of understanding our existence.

In the following sections, we will deal with (1) Japanese positive attitudes toward robots, (2) the relationships between these attitudes and historical and social conditions in Japan since the 1970s, (3) Japanese cultural traditions which are related to the ideas "betweenness," "*aida*" and "*rinri*," and (4) the ethical meanings of robots and robotics seen from Japanese cultural-philosophical perspectives.

2 Robots Seen from Japanese and "Western" Viewpoints

It is often acknowledged that in Japan the discussions on ethical aspects of robots and HRI (human-robot-interaction) are relatively limited, compared to "Western" countries. However, on the other hand Japan is often called the "Robot Kingdom." This might be considered contradictory if seen from a "Western" viewpoint. Therefore, it is necessary to carefully consider the direction of ethical discussions on robots in Japan.

[1] *Aida* as a Japanese term is usually translated into betweenness. But the important point is that this betweenness is not "fixed." *Aida* is "in the process of emerging and changing" or "in the state of interdependency," as we will see in this chapter. For example, according to our understanding, the robot is something related to *aida*. The existence of robots reflects our interpretation on "what is a subject or an object?" And our existence itself depends on this interpretation.

[2] Likewise, *rinri* as another term in Japanese is usually translated into ethics. But the meaning is not fixed in this case either. The meaning of *Rinri* is dependent on the relation between *rin* 倫 and *ri* 理, as we will see later. Generally speaking, *rin* means ethics and *ri* means reason. But *aida* exists between *rin* and *ri*. Think about the case that the robot is something related to *aida* and works as a copula linking *rin* and *ri*. In this case, the robot's work might change the meaning of *rinri*.

Kitano (2006) suggests that there has been little discussion on Roboethics in Japan. This is very strange when we consider the fact that Japan is among the most highly developed industrialized countries and is known as the "Robot Kingdom." Kitano believes the direction of discussions on robots in Japan is more practical and positive than theoretical and philosophical (Kitano 2006: 79).

Krebs (2006) indicates that "Western" people cannot help from feeling "astonishment and alienation" about the positive attitudes toward robots in Japan. Krebs cites various reports on Japanese attitudes toward robots from the "Western" media. For example, Faiola, a journalist shows this feeling in one of his reports in the *Washington Post* (Krebs 2006: 64):

> Though perhaps years away in the United States, this long-awaited, as-seen-on-TV world – think "The Jetsons" or "Blade Runner" – is already unfolding in Japan, with robots now used as receptionists, night watchmen, hospital workers, guides, pets and more. (Faiola 2005)

A reporter in the German news magazine "Die Zeit" uses the term "Land der Robter (Land of Robot)" when informing his audience about this "strange and also ultramodern" situation.

> It is not by chance that Japan is called the "Robot Country." Clever machines have been used here for decades in industrial production and in shopping centers or hospitals, and more recently in schools and universities as well as at home for private lessons. (Lill 2015)

After examining these discourses from "Western" countries, Krebs concludes that Japanese positive attitudes toward robots reflect Japan's cultural traditions such as *Shinto* 神道, polytheism, idea of all-animation, the characteristics of Japan's economic and industrial segments including labor market, employment, labor unions and high education level as well as the positive portrayal of robots in Japan's pop culture such as comic books. In Japan's historical and cultural tradition, the concept of animism can extend itself to animate objects and this contributes to Japanese positive attitudes toward robots.[3]

This tendency is endorsed by some empirical surveys. The author's own "empirical" research ("content analysis") of major newspapers reports, Asahi Shimbun and Nikkei Shimbun, reflect the poverty of ethical and critical discussions on robots in Japan.[4] The data in this research shows that the newspaper articles containing terms such as "Robot & military use," "Robot & ethics" and "Artificial intelligence & ethics" appear seldom. Conversely, terms such as "dream" and "future" appear far more frequently.

Contrary to this tendency in Japan, "Western" people seem to have more critical attitudes toward robots, as the following reports from a "quality paper" as well as a "tabloid paper" indicate. Noel SHARKEY's views on robots are often cited as a typical "Western" case of negative attitudes toward uncritical acceptance of robots in

---

[3] Kreb's discussions on animism as sources for acceptance of robots in Japan agree with Schodt (1988). The parts that correspond to this discussion are on pages 198–200 of Schodt's book.

[4] An analysis of newspaper reports from January 1, 2000 to December 31, 2012 was undertaken. For more details of this analysis see Nakada (2013).

social life. He is a specialist in the research on artificial intelligence and robotics. Sharkey states:

> The deployment of the first armed battlefield robots in Iraq is the latest step on a dangerous path – we are sleepwalking into a brave new world where robots decide who, where and when to kill… The US National Research Council advises "aggressively exploiting the considerable warfighting benefits offered by autonomous vehicles." They are cheap to manufacture, require less personnel and, according to the navy, perform better in complex missions. (Sharkey 2007)

One reporter from the *Daily Mail* endorses this warning, indicating even Bill Gates as Microsoft founder fears robots and AI.

> In the past year experts including Elon MUSK and Stephen HAWKING have said the rise of "super-intelligent" robots poses a threat to humanity. And now Microsoft co-founder Bill GATES, too, has joined the voices calling for caution. He said the rise of AI should be a concern, and he doesn't understand why people are not taking the threat seriously. (O'Callaghan 2015)

3 Factors Affecting Japanese Views of Robots

We can find various reasons behind the positive attitudes toward robots in Japan. One of the reasons can be found in the fact that robots provide Japan's industries, economy and society with hope for the future. The New Energy and Industrial Technology Development Organization (NEDO) explains in the Robot-hakusho in 2014 (*Robot White Paper in 2014*) that Japan needed robots as a new prime mover for transformation of the industrial structure after the oil shock in the 1970s.

> The rise of industrial robots was triggered by technological progress meeting the requirements of the manufacturing industry. Due to the oil shock of the 1970s, Japan's manufacturing industry was forced to change policy from investment for production capacity to improving productivity. The manufacturing industry required automation systems. (NEDO 2014)

The robotics industry has played a role as a symbol of this trend along with high-tech industries such as computers and semiconductors. Importantly, although other high-tech industries such as computers, semiconductors, electric/electronics industry have been losing ground since the 1990s, the robotics industry still remains a premier player in the global market in the 2000s.

This means that the emergence of the robotics industry in the 1970s and the 1980s saved Japan's economy as well as people's lives by suggesting a new direction for economic growth, stabilization and promotion of employment as well as realization of the partnership of humans and robots portrayed in comic books and fiction since the 1950s such as *Tetsuwan Atomu*. Table 8.1 shows this tendency. This table originally appeared in the 2014 NEDO Robot White Paper and has been adapted by the author.

The new direction for the Japanese robotics industry discussed by the government, robotics experts and private enterprises in recent years concerns social robots.

Table 8.1 The number of shipments for industrial robots of major countries, Years 2003–2012

	Year 2003	2005	2007	2009	2011	2012
Japan	31,588	43,922	36,091	12,767	27,894	28,680
USA	11,286	17,677	14,978	6831	20,555	22,414
Germany	13,081	10,075	14,721	8507	19,533	17,528
France	3117	3007	2736	1450	3068	2956
China	1451	4461	6581	5525	22,577	22,987
South Korea	4660	13,005	9048	7839	25,536	19,424
Thailand	156	1458	1252	774	3453	4028

Adapted from the 2014 *NEDO Robot White Paper* which was compiled from the International Federation of Robotics: *World Robotics 2013*

The Ministry of Economy, Trade and Industry (METI) compiled estimates of the size of the robotics industry market in the future in Japan (METI 2013).[5] It seems that METI has a strong hope for the expansion of Japan's robot market and service robots are expected to play a central role of this market in Japan. However, considered objectively, this is regarded as an overestimation. Hobby robot sales are expected to reach 1485 billion yen in 2025, but were just 1~6 billion yen in 2012. Most hobby robots are very expensive. The seal robot *Palo* (Daiwa House Industry) is 420,000 yen (in 2010). Sony's *Aibo* was around 200,000 yen. Mitsubishi Heavy Industry's *Wakamaru* was 1,575,000 yen.

Japan' expectations for service robots are meaningful when seen from ethical viewpoints. METI's estimation of the size of the robotics industry in the future does not include military applications. This view is different from other countries. Japanese people don't worry about use of military robots by their country and society in the future. The meaning of "service" of service robots in Japan is usually limited to consumer products, welfare or amusement.

4 Cultural and Ethical Traditions Behind Japanese Attitudes Toward Nature, Tools and Machines

What we can learn from the discussions mentioned above is the importance that Japanese robotics reflects values, historical process and future images of Japanese life with robots as well as economic and industrial politics. In this sense, contrary to what we might expect, Japanese people live in a world where the ethical problems of robots are important. In fact, people have paid a lot of money in tax for development of various robots. According to a Forbes' report, *Paro* a pet robot needed 12 years and $10 million as time and cost for development (Rose 2005).[6] Besides, even if *Paro* plays an important role in an ageing society, there still remains the

[5] METI's estimation is partly based on NEDO's 2010 survey.

[6] Lacey Rose is a Forbes staff member.

question of whether an animal-like machine is a real substitute for "true" animals.[7] (This means that in Japan people face a set of ethical problems with regard to communication with a pet robot although they are not aware of this ethical situation so clearly.)

We don't know what kind of situations would occur as a result of the interaction of humans and humanlike machines. Robots as humanlike machines would emerge through artificial integration of various humanlike functions. Robots need sensors, computers and motors or artificial eyes, artificial brains and artificial muscles. This set of functions is related to definitions of robots.[8] This means that robots are fundamentally characterized by differentiation of various functions and senses. On the other hand, some scholars say that human infants are in a state of un-differentiation, that is "Sounds trigger both auditory and visual and tactile experiences (Baron-Cohen 1996).[9]" Robots move toward artificial un-differentiation but humans move toward differentiation during the process of development.

In this sense, Japanese people need ethical discussions on robots. To understand robots is to know their life, society and existence. The following authors and scholars show us material to discuss the reasons why "Japanese did not feel aversion to the "mechanical monsters" common in the West" (Krebs 2006).[10] KITANO Naho 北野菜穂 thinks that two cultural factors are important with regard to Japanese attitudes toward robots; animism and Japanese ethics, *rinri*. According to Kitano, in the traditional culture grounded on Japanese mythology and indigenous religion (*Shinto*), there is a belief that the spirit (*tama* 魂) exists in objects and natural phenomena. Even in later periods, this belief remains and expands to include artificial objects. In this tradition, various objects, phenomena, utensils, tools, even artificial objects are believed to have their own spirits and these spirits are thought to be in harmony with human beings. In addition to this kind of belief or value, Japanese have another belief that "*rinri*" (for "ethics") comes from a person's position in this world and community. A person's existence is inseparable from their coexistence with other people and their social responsibility. This concept of "*rinri*" is different

[7] Forbes's report gives a positive evaluation on *Paro* as "newer robots have the potential to serve as much more than companions." "The machines could monitor aged patients, watching out for falls, and remind them to take their medications. Additionally, they could serve as communication tools …" However it adds: "But experts are not suggesting that robots will replace live pets altogether" (Rose 2005).

[8] The definition of robot by ISO-Standard 8373 is: "A robot is an actuated mechanism programmable in two or more axes with a degree of autonomy, moving within its environment, to perform intended tasks. Autonomy in this context means the ability to perform intended tasks based on current state and sensing, without human intervention." This definition is cited on IFR's website. Available from http://www.ifr.org/service-robots/. Accessed 18 November 2015.

[9] In this paper, Baron-Cohen shows us that this "un-differentiation" is related to the hypothesis of the Neonatal Synaesthesia. This is in contrast with the Cross-Modal Transfer hypothesis emphasizing the interaction of modalities in the early stages of infant development.

[10] Krebs cites two reports (Faiola 2005, Wagner 2005) concerning this difference between Japan and the "West." He points out that "feeling aversion to humanoid robots" is a typical attitude of "Western" people.

from the Western concept of ethics as more subjective and individualistic (Kitano 2006: 79–82).

Kitano's explanation of *Rinri* is adapted from the ideas by Japanese philosopher and ethicist, WATSUJI Tetsurō (1889–1960). Kitano adds:

> if we can combine animism with Watsuji's idea of *rinri*, we can find a potential direction for ethical discussions on robots in Japan. This is because animism can be interpreted as a kind of "betweenness" which suggests that there can be *rinri*-based relationships between humans and things including robots as well as nature, things, utensils and tools. According to Kitano, we need to examine the "betweenness" of human beings and as the next step find the "betweenness" of an individual human being (the robot owners) and the spirit of robots as mechanical objects. Kitano explains that Watsuji's work starts from study of *ningen* 人間 (human beings). *Ningen* is composed of two characters, the first, *nin* 人, meaning "human being" or "person," and the second, *gen* 間, meaning "space" or "between." Thus, *ningen* literally reflects the connotation of existence in the "betweenness." The essential point of this idea is that human beings are not isolated entities but entities fundamentally associated with mutual relations and mutual respect.[11] Human beings as *ningen* are characterized by individuality (*nin*) and social beings (*gen*). (Kitano 2006: 80–82)

We need to add further explanation to Kitano's discussions to understand "betweenness" (or *aida* in Japanese) and apply this concept to potential discussions on robots in Japan. "Betweenness" is usually understood to reflect the relationship between humans, not between humans and things. Kitano's interpretation of this point is not clear. One of the important points is that Watsuji's discussions might not be reduced to mere subjective attitudes toward robots and tools. Kitano suggests:

> As far as the owner treats the robot (or the spirit of robot) with a proper manner, the robot should have the respect for the owner, act with harmonization, and ethical behaviour. Thus spatially, the togetherness of the existences of the human (the owner) and the robot (the tool) constructs the range of their betweenness. This belief is reinforced by the idea of animism as the robot is able to have its identification only while the owner is using it. (Kitano 2006: 82)

However, this sounds too subjective in the sense that *rinri*-related attitudes emerge only within human beings. It seems that we need our own interpretation of Watsuji's *rinri*. It is paramount to determine the necessity for *rinri*. This is probably because we are in the state of plurality of individuality and sociality. Human beings are characterized by the tension of two states of being, that is as a *hito* 人 or person (*nin* = *hito* = person) and in-between-ness or betweenness (*gen* 間 = *aida* = space = relation). In this sense, *rinri* means the understanding of this state (*ri* = reason = understanding) and also the requirement necessary for deeds toward *rinri*-based relations (*rin* = pursuit for virtue or authentic ways of life). (This understanding of *rinri* reflects the author's own interpretation. *Rinri* requires us to behave as humans following *rinri*. The understanding and the praxis are not separated from each other in this case.)

---

[11] Kitano explains that the meanings of Japanese *rinri* translated as "ethics" which appeared in Japanese society and culture in 1881 but Japanese people used the term *rinri* before this import of "ethics" from "Western" cultures (Kitano 2006: 80–82).

This point is important and difficult. Therefore, the authors and scholars who indicate this plurality are limited. The article about Watsuji in the *Stanford Encyclopedia of Philosophy*[12] is one of the exceptions. It indicates:

> At the center of Watsuji's study of Japanese ethics is his analysis of the human person, in Japanese, ningen… Ningen is composed of two characters, nin, meaning "person" or "human being," and gen, meaning "space" or "between." He cautions that it is imperative to recognize that a human being is not just an individual, but is also a member of many social groupings. We are individuals, and yet we are not just individuals, for we are also social beings; and we are social beings, but we are not just social beings, for we are also individuals. Many who interpret Watsuji forget the importance which he gave to this balanced and dual-nature of a human being.[13]

This interpretation on "double negation" as characteristics of human existence sounds interesting. It indicates that we need *rinri* because of this negation. Negation also means distinction. Human beings are "divided from the originally unified self,[14]" therefore we can understand that we are apart from the original unification or oneness. There is distinction or distance between us and original unification. However, this distinction or distance provides space to enable us to see why we need *rinri* as a combination of reason and practice. *Rinri* is important for us human beings who know that we are determined to be individual but also know the origins of this isolated individuality at the same time.[15]

Our understanding is far better than previously but still limited. We still need additional interpretation of "why *rinri* can exist for robots or for human-robot-interaction."

It is often said that "Western" people are surprised to know that Japanese give human names to even industrial robots. This difference reflects cultural differences between "Western" culture(s) and Japanese cultural tradition. Positive Japanese attitudes toward robots can be explained as a reflection of animism, *Shinto* and Buddhism. This is what Kitano (2006) intimated and similar remarks can be found in various papers, books and news reports. The following is a typical case of these remarks by one of the most well-known researchers on robotics in Japan.

> Japanese treat anything in the universe as if it has a soul inside… We cannot treat them less worthily (rudely / roughly /impolitely) or even too-worthily (too-goodly / too-muchly) because we are no more than they are and even some of them become a god. We say Mottainai if someone treats them rudely /roughly /impolitely or even too-goodly/too-

[12] The *Stanford Encyclopedia of Philosophy's* entry on "Watsuji Tetsurô." (First published on November 11, 2004; substantive revision appeared on October 7, 2014.) Available from http://plato.stanford.edu/entries/watsuji-tetsuro/. Accessed 30 November 2015.

[13] Stanford Encyclopedia of Philosophy's entry on "Watsuji Tetsurô."

[14] Stanford Encyclopedia of Philosophy's entry on "Watsuji Tetsurô."

[15] On this point, we remember NISHIDA Kitarō's "absolutely contradictory self-identity" (*zettai mujunteki jikodōitsu* 絶対矛盾的自己同一). In one of his essays, Nishida explains very simply: "we humans have the potentiality to violate the law and therefore we have our self. Our self in the sense of praxis is not understood by the mere reason" (Nishida 1989: 332). We believe that this is also a double negation but this can't be separated from our praxis, i.e. to know our self through contradictions and decision.

> muchly. Mottainai is something much more than Waste to Japanese which makes the society to be highly ecological and highly friendly to anything including artificial ones.[16]

Selma Šabanović who has been working with Japanese roboticists agrees with this remark, saying:

> Robotics researchers refer to animistic beliefs and practices to suggest that Japanese people are particularly susceptible to interpreting robots as companions and to legitimize the creation of robots that perpetuate an anthropomorphic view of technology. (Šabanović 2014)

According to this explanation, the preference for humanoid robots by Japanese roboticists reflects their belief in the uniqueness of Japanese cultural tendencies. These preferences were determined through interviews with Japanese roboticists at Waseda University or AIST. As a result of this preference they have the tendency to create humanoid robots which are believed to be suitable embodying *kokoro*[17] into robots. For them *kokoro* is more important for robots and human beings themselves and this is in contrast with the definition of intelligence as rationality in Western concepts. For them *kokoro* as a holistic concept is the integration of emotion, intelligence, and intention (Šabanović 2014).

However, it is not clear whether these kinds of attitudes toward robots can be found only in Japan and not in other countries. In fact, Leander Kahney reports that more than half the owners of iRobot's *Roomba* name their device. Owners often talk to their machines, and many treat them as though they were alive, or semi-sentient. Some even take them on holiday, unwilling to leave them at home alone. *Roomba* is a robot vacuum cleaner and different from robot pets such as Sony's *Aibo*. But *Romba*'s owners "acquire similar feelings for their robot vacuum cleaners," just as *Aibo* "develop[s] emotional attachments to their mechanical companions." Similar things happen in the case of the *Trilobite* robovac in Europe (Kahney 2003).

MIT anthropologist Sherry Turkle's discussions on the *Tamagotchis* focus on how these relational artifacts are similar to and different from toys like teddy bears, rag dolls, and other inanimate objects. The similarity is they are "alive" or "sort of alive" for children at a certain stage of development. The difference is, for example, that we do not know whether the *Tamagotchis* can become another "transitional." As D.W. Winnicott shows, children need teddy bears or dolls as "transitional" objects. As transitional objects, they mediate between children's inseparable mother and other people as separable beings and also between inseparable parts of the self and beings as "not me" (Turkle 2007, 2011; Winnicott 1971).

In this sense, the remarks about Japanese robots which are supposedly typical of Japanese thinking about robots might be a kind of an unverified hypothesis. This does not mean that humans' empathy for robots and *rinri* for robots based on this kind of empathy or "in-between-ness" between different entities and beings as

[16] Slides from Takanishi Atsuo (2007). Takanishi's remarks on this subject are shown in the following article: Šabanović (2014). Selma Šabanović is a Professor at the School of Informatics and Computing, Indiana University, Bloomington, USA.

[17] This term *kokoro* is usually a Japanese equivalent of "mind" but as Šabanović's explanation shows, it is regarded to include more connotations and denotations in some cases.

human-like machines and humans are not impossible or important. Some researchers provide evidence about the potential "in-between-ness" between humans and robots. Suzuki et al. (2015)'s research shows that humans have the ability to empathize with robot's pain under certain circumstances. They found that the electroencephalographic data from human subjects show differences in two cases; one case showing a knife on a human or a robot finger (the painful situation) and another case showing a knife near but not on the finger (the non-painful situation). Interestingly, the human subjects show responses to the painful situation of a robot too. The researchers interpret this finding as evidence of humans' ability to empathize with the robot's pain.

In this sense, *rinri* or *rinri* for robots provides an important suggestion on how and in what way a robot can be an ethical being. As we have partly examined, this subject cannot be answered as easily as most Japanese roboticists might imagine. This is because a robot is a composed entity consisting of technological, ethical and historical aspects.

5 Spatiality and *Rinri* for Humans and Robots

The next part of this chapter will discuss the realization of *rinri* as ethical attitudes toward robots while examining the meanings of Japanese ethics and philosophy related to potential roboethics in Japan. This also concerns "naming a robot" (assigning a human name to a robot) typical of Japanese attitudes towards a robot.[18] When some workers in factories name their industrial robots in their working units "*Momoe*" or "*Ikuyo*," typical human names in Japan, "betweenness" emerges. Or at least this naming is performed with the explicit or implicit presupposition that human beings exist in the space or the place called "*gen*," "*aida* (betweenness)" of Watsuji.

In his "*Fūdo* (Climate)" Watsuji contrasts his idea of "space" with Heidegger's idea of "time" or "temporality" in *Sein und Zeit* (*Being and Time*). Watsuji says that Heidegger's ideas of time and Dasein are grounded on isolated existence and therefore his discussions on human existence appear within isolated consciousness. On the other hand, within the space, "betweeness" or "*aida*" is the primordial ways of being (Watsuji 1979).[19]

Watsuji suggests that, in this space, interactions and relationships are the sources for meanings of various entities and beings. Within the space there is no transcendent

[18] Naming a robot is related to *aida*. When we name a robot as something like a human, this naming will change some aspects of a robot. A robot will enter our life space as an individual existence with a name. And at the same time we know that the robot is a species-being or genus-being too. The robot doesn't have a name as this kind of being. Various problems regarding robots might occur from this dualism or *aida*. When a robot with a name or without a name takes care of us, are we beings with a name or without a name?

[19] For this contrast, see pages 3–4 of this book. *Fūdo* was originally published in 1935.

Other. Without the transcendent or absolute other, in the space, mutual relations and "reciprocity" create the conditions for a union between humans and humans as well as humans and things. Within this climate, human beings create tools, houses and clothes as objectified (externalized) others. These things are objectified and become the conditions for objectification of human beings. This is because humans experience themselves through these objects. We experience "coldness" via clothes and "heat and humidity" through use of a small fan in our hand in the summer and growing rice in the rice field. Our experience of ourselves and therefore our understanding of ourselves can be made possible through objectified (externalized) objects. In this sense, the relationship between a person and an object in climate or in space are interrelated (Watsuji 1979: 24–25).

Tomiyama (2004) explains that according to Watsuji's "*Rinri-gaku* 倫理学 (Ethics)" (Watsuji 1937) objects can have duality. Objects are objectified as tools but at the same time these objects can be regarded as "subjective" or "active" entities because they work as mediators for human existence and humans' self-realization. In Watsuji's discussions on the duality of objects, we can find suggestions to understand this space as "the tracks" of development of our subjectivity.[20]

YUASA Yasuo, Watsuji's former student, says that Watsuji's discussions on space and NISHIDA Kitarō's concept of *basho* 場所 (place) contrast with the tradition of Western philosophy. In Western views, the emphasis is on "time" and "mind," and in Japanese views the emphasis is on "space" and "body" (or undifferentiated status of mind and body). Yuasa believes the tendency of putting emphasis on mind and time rather than on body and space was established by Christian views of the world and humans, in particular, at the time of Augustine. The essence of these views is that the human mind can hold the flow of time, divided by divine creation and eschaton (end of time) within oneself and this enables the human to leave one's life in the space and one's corpus (body) and meet God within his anima (mind). The human mind is the place where the spiritual nature of *imago Dei* (the image of God) dwells and the corpus with original sin dwells in the space. Yuasa writes that Descartes' theory of mind-body dualism reflects this divine view in the sense that the essential point of human existence is characterized by one's mind. On the other hand, for Watsuji and Nishida, the primary matter is that our body exists in space (Yuasa 1990).

The discussions mentioned above are important. However, some difficult problems still exist, including the problem, of why or in what ways humans enter the *rinri*-based relationships with robots. The central points of these discussions are that the undifferentiated situation of mind and body and also the unity of a thing and a person as the primordial ways of existence for various entities and beings can emerge in the space called "betweeness" or "*aida*." Without doubt Japanese philosophers and ethicists provide us with various suggestions on these points. For

[20] Tomiyama tries to combine the topic of "modern subject" with Watsuji's ideas of "*aida*" or "*aidagara*" (relations in the space or *aida*) in order to solve the problems of human relations. He thinks that Watsuji's "*aida*" can be interpreted as a kind of "communication theory" which is meaningful in the field of media studies.

example, Nishida's pure experience (Nishida 1950) and active intuition and Kimura's *koto* こと (phenomena and events based on *aida*) and *mono* もの (things) (Kimura 1994) provide useful examples of the combination of Japanese ways of life, cultural traditions, language and ethical thoughts.

However, the problem still remains of how and in what ways robots are different from other *mono.* This is crucial point when considering the nature of ethical relationships between humans and robots and what humans can learn about themselves from the relationships between humans and robots. If human beings and robots are in a state of reciprocity or "betweenness," it is possible to determine the meaning of our existence from this reciprocity or undifferentiation.

6 Human and Robot Forms

ICHIKAWA Hiroshi's ideas of "body" provide points to consider among other matters how and in what ways robots are different from other *mono.* The important point is that "undifferentiation" can be interpreted as active duality.

Ichikawa tries to combine the Japanese senses of *mi* 身 (body) with Merleau-Ponty's mind-body theory, neuroscience, biology and ethology. *Mi* is usually translated as "body" but this translation lacks important nuance. Ichikawa explains that *mi* is a body as a mind and also *mi* is a mind as a body. However, this explanation is limited. There is an expression "*Mi wo noridasu* 身を乗り出す" which can be translated as "lean forward." In English some important nuance is lacking here too. When we say "*Mi wo noridasu*," this shows a body pose which is to take one step forward towards the matter or situation of interest. This body pose simultaneously shows our mind which is engaged in this interesting situation. As this example shows, *mi* has two meanings related to body and mind at the same time. Ichikawa says that human beings with *mi* exist in a world which has distance and perspective. Furthermore, this kind of world for humans with *mi* is not a world which includes God as a manifestation of the mind or a purely mental reflection. Being a purely mental reflection means that distance and perspective are not needed and not-purity as characteristics of human with *mi* is out of consciousness (Ichikawa 1992).[21]

Ichikawa adds that our body as a body-mind complexity is the place where subjectivity-objectivity, activity-passivity, to see-to be seen and *mono*-person interrelate. When something touches our hand, the relationship between being passively touched and actively touching is not as clear as we usually imagine. This is less clear when we touch our hands using our right hand and left hand. The relationship between to be touched and to touch is reciprocal. To touch and to be touched merge. The phenomenon of personification or empathy with things as others is grounded on this ambiguity or reciprocity (Ichikawa 1992: 102).

Ichikawa indicates the characteristics of the object which humans as the subject relate to as "hard," "soft" or so on, reflect the "softness" or "hardness" of our own

---

[21] Here the author added his own explanation about "*Mi wo noridasu*" to Ichikawa's concept of *mi*.

objectified body too. This means that we can deepen our own understanding through deepening our inner understanding of objectiveness of the object. In this sense, "to feel and to be felt" mirrors the state of reciprocity. Similarly, on a deeper level "to see and to be seen" reflects a state of reciprocity also. In particular, our body is important as something to see and something to be seen (Ichikawa 1992: 86–91).

Yuasa Yasuo believes that in a Noh play undifferentiation can be achieved as a form of the Noh player's performance. In a Noh play, the aim is to form a dramatic posture or a dramatic pattern of movement. The meaning of the dramatic movement in Noh is described in essays by Zeami 世阿弥 (1363–1443). As an unskilled practitioner the body of a Noh player is similar to an objectified thing which the Noh player cannot control very well. The mind of the player is unable to control the body. Sooner or later the player moves to a stage of proficiency, whereby the player's mind reaches a stage of *mu* 無 (nothingness) or *mushin* 無心 (mindlessness). The body can move freely following the acquired pattern of movement. At this stage the body does not need direction from the mind. Thus the mind becomes *Mu*. Or we might say that the pattern of the body movement becomes the mind itself (Yuasa 1990: 138–139). At this stage a linkage of the body-mind-pattern emerges.

Therefore, the space mentioned by Watsuji is the place where this linkage emerges or this linkage enables the space as "beweenness" to emerge. In addition, the relationship between the player and the audience can be added to this linkage of body-mind-pattern. The pattern of movement of the Noh player presupposes the presence of the audience. The player's movement makes the pattern something to be shared by the player and the audience. Further, in the relationship between "form of movement of the player" and "expression," this linkage emerges as the complex link of body-mind-pattern-the player-the audience-expression.

The important point regarding this linkage is in the process and dynamics. This linkage is not fixed. This can be seen from various viewpoints, example, that of Nishida Kitarō. In Noh, the pattern is also the form. The form or the pattern reflects the state of duality. The form or the pattern of stylized and dramatic movement shows the mind and at the same time the body as an objectified (externalized) object. If we use Nishida's terms, this pattern or the form of movement or the linkage of form-movement-body-mind-expression is also the fusion of "action and intuition" and "poiesis and praxis." The stylized movement of the Noh player is an expression. This expression is in the state of duality. This work is an objectified (externalized) object and also reflects the inner mind of the player. The movement is action and also seeing. So the movement is in the duality of action and intuition, or a sort of action-intuition. The stylized movement is the "praxis," a contribution toward the world and this "praxis" is "poiesis" (to mold something) at the same time. The term Nishida uses "*kōiteki-chokkan* 行為的直観" can be translated as "action-intuition," "active intuition" or "enactive intuition." This term signifies the unification of the subject and the object and also the intuition and the action. The assumption is that action and subject are active and intuition and object are passive. However, this difference collapses in this situation or *basho* (place or topos) and the distinction between "praxis" and "poiesis" disappears. Nishida states "Our action is always an expressive one. Seeing is forming and forming is seeing. The subject determines the object and the creation of objects is done as a form of unification of the subject and the object."[22] (Nishida 1935: 145).

This kind of orientation to undifferentiation or reciprocity can be seen as characteristics of Japanese attitudes toward tools, utensils or aesthetic experiences. Although further deepening of ideas about *rinri*, *aida* or undifferentiation is necessary, it is clear that we have moved further from the previous viewpoints constrained by isolated entities or mind-body dualism. What we can imagine is that isolated entities based on mind-body division are related to "isolated sense of ethics." We don't know how the oneness of body and mind or humans and other entities can make a different kind of ethics emerge. But we can certainly imagine something different after we think about the phenomena mentioned here. Ethics is not separated from our effort to know this world and our existence deeply.

7 Robots in Japanese *Aida* or Spatiality

From the perspectives related to this orientation of undifferentiation or reciprocity, robots can be viewed differently from the "Western" views grounded on differentiation of body and mind and objects and subjects. The definition of a robot cannot be separated from its design. The design of a robot can be seen as a certain type of expression about what a robot is and what a robot can do and this expression takes the form of a human shape and mechanical devices. These elements cannot be separated from each other. This unification can be made understandable through use of adjectives, for example the adjectives used when we use a tool or a machine moves. These include adjectives such as "this is human-like," "this is useful" or "this is heavy."

This reflects Heidegger's remark "this hammer is heavy" when he explains the difference between "readiness-to-hand" (*Zuhandenheit*) and "presence-at-hand" (*Vorhandenheit*):

> the less we just stare at the hammer-Thing, and the more we seize hold of it and use it, the more primordial does our relationship to it become and the more unveiledly is it encountered as that which it is – as equipment. (Heidegger 1962: 98)

This reflects also Nishida's pure experience. The important thing is that the tools, machines and technology used for these things can represent an utterance such as "this is human like" or so. At the same time, these things depend upon this kind of utterance or use of adjectives. This is the basis of and what can be learned from Japanese perspectives on undifferentiation or reciprocity.

As we have discussed citing Watsuji and Yuasa, we understand ourselves by the objectified (externalized) objects. The objectified objects enter our life via the expressed body pose (with adjectives describing this pose). And now we interact with various artificial objects in our digitalized environments. So to examine the

[22] English translation by the author (Nakada).

artificial objects in our new environments is an important subject for us. The author will do this effort in the following passages.

Honda's website shows us that we can know the mechanism of robots' and human walking, comparing these different bodily entities. We know how our body works and how balance of our bodily parts can be maintained when we walk through observation of robots. In the case of robots such as *Asimo* using two legs, the balance of bodily parts and two competing phases of movement, maintaining a standing position and making a foot step forward, is determined by the relationships of the "floor reaction controller," "position controller" and "target ZMP (Zero Moment Point) controller."

These three forces are "keep the foot firmly on uneven ground," "lean the upper body forward to keep a standing position when a robot's foot can't work to maintain balance" and "control the position of the upper body caused by target ZMP by changing stride."[23]

This is a matter of technological mechanism. However, the sources of this mechanism come from the mechanics of the human body and we know about our bodily mechanics through this technologically constructed robot body. This mirrors the relationship between mind and thing mentioned above. Take, for instance, the case of a bipedal robot. The bipedal robot requires a human like body. What is a human body can be seen through research on the walking pattern of mechanical human-like legs. The walking pattern of a robot is not natural during the first stage but through research on this "not-natural walking pattern," a robot can move more naturally.

In the case of the form of movement by an expert Noh player, the mind disappears as a result of practice. The form of dramatic movement expresses the mind of the player itself. In the case of the robot with a "not-natural walking pattern," the mind re-appears as a kind of a question of "why the robot can't walk naturally." In this sense, the meaning of "natural" or "unnatural" can't be separated from human bodily existence and vice versa, we can know (at least partly) "what is to be natural" and "what is a human body" through research on robots.

The difference between a robot using "passive dynamic walking" and a robot under the control of a "ZMP controller" is a difference of purely technological matters. These technological matters cannot be separated from subjective understanding of the ways a robot walks which can be explained using such adjectives as "efficient," "passive," "dynamical" or "balance of to be passive and to be dynamical (Ura et al. 2015)." This can be interpreted as a case where robots learn from the bodily structure and bodily existence of humans and at the same time humans learn from robots. When we walk, we are not aware of the mechanics of walking. In this situation our mind does not control our body intentionally and consciously.

In this sense, roboethics is a subject to be examined from "betweenness" perspectives and our human existence as mind-body unity is a subject to be examined through observation of "*aida*"-relations, mind-body interactions and meanings of "betweenness" in specific situations. Furthermore, the meanings of "*aida*" or

[23]Available from: http://www.honda.co.jp/factbook/robot/asimo/200011/04.html. Accessed 6 December 2015.

"betweenness" can change by the emergence of a new kind of interaction between humans and artificial objects as well as entities in real and virtual situations.

Therefore, roboethics and research on "*aida*" or "betweenness" itself are also in a state of reciprocity. The meanings of ethical studies on robot and our pursuit of "*aida*" (betweenness) existence are dependent on each other. Research on "O-space" provides us with ideal subjects on this point. According to Adam KENDON (Kendon 1990, 2010[24]), when we are engaged in face-to-face interactions between two or more participants, the standing positions of the participants form some sort of spatial organization called F-formation (or *facing formation*). This formation forms another space called an O-space. This O-space is a transactional space made by the participants' transactional segments which are overlapping. As Edward HALL suggests, we have various types of space between other participants in transaction or communication and this space determines the transaction and communication process (Hall 1966). Yamaoka et al. (2009) suggest that O-space is the space where the participants can talk about common topics and see interesting objects. The important point regarding O-space is that this spatiality exerts an influence on the transaction and communication between a person and a robot too.

According to Kanda (2008) who reports on people's reaction to a communication robot on display in a museum of science, people's reactions vary with the distance from the robot. People stop around the robot and watch it at a distance of 2 meters. At a distance of 1.25 meters people talk to a robot and at a distance of 0.75 meters physical contact and conversation begin. This tendency is similar to a transaction between people. This spatiality influences the behavior of a robot too. A robot might surpass humans in speed of information processing and amount of stored information within it. However, as the cases mentioned above show, what a robot can do or talk about is dependent upon O-space which includes transaction segments from both the robot and human side.

According to psychological research on aphasia, patients with aphasia seem to have difficulty in treating information or objects within the O-space. Within the O-space, "this" for "me" becomes "that" for the other participant. This is the exchange of viewpoints. The difficulty emerges with regard to this exchange of viewpoints in the case of aphasia (Kumagai 2011: 26–27). In this sense, *aida* refers to pluratiy which includes the space of objects, the interaction of objects, human inner mind and human intersubjectivity. The case of aphasia suggests us this pluratiy as one of important human conditions.

In Japanese there is an expression "*ano*" (that time or place for us) *toki* (time). "*Ano*" is different from "this" and "that." "*Ano toki*" refers to a certain time in the past and the time when the participants can share memories or imagine their common presence there and then. So in this sense, "*ano*" refers to a different O-space which is not present here and now but is thought or is imagined to be somewhere in the past.

---

[24] The author used Kendon (2010) to summarize the concept of O-space.

This can be interpreted as the source of some kind of time. "*Ano-toki*" can emerge only through a certain kind of O-space. So we might say that spatiality and time are also in a state of reciprocity.

8 Conclusions

The typical topics of roboethics in the "West" such as "autonomy" and "responsibility (of robots)"[25] reflect the cultural and philosophical traditions of the "West" which emphasize the priority of mind and time over body and spatiality.

These topics appear differently when seen from Japanese viewpoints. For example, the phenomena of fusion of different modes of senses cannot be visible from the perspectives grounded on mind (separated from body or from unification of mind-body) and time or on the presuppositions of differentiation of mind and body.

Research by Iriki et al. (1996) and Maravita and Iriki (2004) provide us with good examples to be considered in this way. Their research shows that the human body and a thing merge into each other under certain circumstances. When humans or apes use objects as tools, the related (intraparietal) bimodal neurons change in such a way to include the fusion of a thing and the body in the schema of related bimodal structure. When we use a long stick as a walking stick, we feel as if our extended body touches the ground directly. This is the fusion of our body and a thing. This proposes important questions about whether a similar phenomenon would occur between us and a robot. In this instance it is necessary to determine what would take place between robot's body and a thing. These are interesting and important ethical questions and can be answered only through use of spatiality. The fusion of different entities seems to need a body or spatiality.

It is necessary to think about the fusion of time (history) and spatiality also. As discussed previously, the need for robots derives from historical situations. This means that the spatiality or *aida* depends upon history. *Aida* is under the influence of past history or encounters with the "West." *Aida* is not fixed. In this sense, it is necessary to consider spatiality in time also.

Watsuji's *rinri* and Nishida's thought of *basho* are not closed. The thing and the mind are in the state of interdependency and this way of interdependence prescribes and reflects the way of the space and time in which we live. We experience that the robot is entering our place and time. This means that Japanese people have started to live in a different way of place and time. This will also be the same for those living in "the West." And this needs a different way to understand our place and time than we used to do so. According to Watsuji, thought and practice needs oneness. This article also reflects this idea on condition that we could successfully enter into this duality, thinking and practice, a thing and a mind.

[25] Concerning the tendency of ethical discussions on robots in the "West," see: Asaro (2007). Floridi and Sanders (2004), Nakada (2010, 2013), Nakada and Capurro (2013), Sullins (2007), and Veruggio and Operto (2006).

References

Asaro, Peter M. 2007. "Robots and Responsibility from a Legal Perspective." Paper presented at ICRA'07: 2007 IEEE International Conference on Robotics and Automation, April 10–14, at Angelicum University, Rome, Italy.

Baron-Cohen, Simon. 1996. "Is there a normal phase of synaesthesia in development?" *Psyche* 2.27: 223–228.

Faiola, Anthony. 2005. "Humanoids with Attitude." *Washington Post*, 11 March 2005. http://www.washingtonpost.com/wpdyn/articles/A25394-2005Mar10.html. Accessed 18 November 2015.

Floridi, Luciano, and J. W. Sanders. 2004. "On the Morality of Artificial Agents." *Minds and Machines* 14.3: 349–379.

Hall, Edward T. 1966. *The Hidden Dimension: Man's Use of Space in Public and Private*. London: The Bodley Head.

Heidegger, Martin. 1962. *Being and Time*, translated by John Macquarrie and Edward Robinson. Oxford UK: Blackwell.

Ichikawa, Hiroshi 市川浩. 1992. *The Body as Spirit* 精神としての身体. Tokyo 東京: Kōdansha 講談社.

Iriki, Atsushi, Michio Tanaka, and Yoshiaki Iwamura. 1996. "Coding of Modified Body Schema during Tool Use by Macaque Postcentral Neurons." *Neuroreport* 7.14: 2325–2330.

Kahney, Leander. 2003. "The New Pet Craze: Robovacs." *Wired* (online), 11.06(June). http://archive.wired.com/science/discoveries/news/2003/06/59249?currentPage=all. Accessed 6 December 2015.

Kanda, Takayuki 神田崇行. 2008. "The Distance of Communication Robot and Human コミュニケーションロボットと人間との距離." *IPSJ Magazine* 情報処理 49.1: 24–29.

Kendon, Adam. 1990. *Conducting Interaction: Patterns of Behavior in Focused Encounters*. Cambridge: Cambridge University Press.

Kendon, Adam. 2010. "Spacing and Orientation in Co-present Interaction." In *COST 2102 Int. Training School 2009, LNCS 5967*, edited by Anna Esposito et al.: 1–15. Berlin: Springer.

Kimura, Bin 木村敏. 1994. *Considering the Pathology of Mind* 心の病理を考える. Tokyo 東京: Iwanami Shoten 岩波書店.

Kitano, Naho. 2006. "'Rinri': An Incitement towards the Existence of Robots in Japanese Society." *International Review of Information Ethics* 6: 78–83.

Krebs, Stefan. 2006. "On the Anticipation of Ethical Conflicts between Humans and Robots in Japanese Mangas." *International Review of Information Ethics* 6: 63–68.

Kumagai, Takayuki 熊谷高幸. 2011. *Japanese as a Visual Language* 日本語は映像的である-心理学から見えてくる日本語のしくみ. Tokyo 東京: Shinyosha 新曜社.

Lill, Felix. 2015. Der bessere Lehrer. *Die Zeit* (online), 24 September 2015. http://www.zeit.de/2015/37/roboter-lehrer-schulen-japan. Accessed 18 November 2015.

Maravita, Angelo, and Atsushi Iriki. 2004. "Tools for the body (schema)." *TRENDS in Cognitive Science* 8.2: 79–86.

METI. 2013. *Trends in the Market for the Robot Industry in 2012*. http://www.meti.go.jp/english/press/2013/0718_02.html. Accessed 18 November 2015.

Nakada, Makoto. 2010. "Different Discussions on Roboethics and Information Ethics Based on Different Cultural Contexts (*BA*): Discussions on Robots, Informatics and Life in the Information Era in Japanese Bulletin Board Forums and Mass Media." In *Cultural Attitudes towards Technology and Communication 2010*, edited by F. Sudweeks, H. Hrachovec, and C. Ess: 300–315. Murdoch: Murdoch University.

Nakada, Makoto. 2013. "How do Japanese people and 'Western' people understand the meanings of 'autonomous' robots and AI in their lives?" Paper presented at CEPE 2013, July 1–3, at Autónoma University, Lisbon, Portugal (*Proceedings of CEPE2013*:101–114).

Nakada, Makoto, and Rafael Capurro. 2013. An Intercultural Dialogue on Roboethics. In *The Quest for Information Ethics and Roboethics in East and West*, vol.1, edited by Makoto Nakada, and Rafael Capurro: 13–22.

NEDO (New Energy and Industrial Technology Development Organization) 2014. *NEDO-Robot-hakusyo 2014* (NEDO-Robot-Whitepaper 2014). http://www.nedo.go.jp/library/robot_hakusyo.html. Accessed 18 November 2015.

Nishida, Kitarō 西田幾多郎. 1935. "The Standpoint of Active Intuition 行為的直観の立場." In *Complete Works of Nishida Kitarō* 西田幾多郎全集. Vol. 8: 107–218. Tokyo 東京: Iwanami Shoten 岩波書店.

Nishida, Kitarō 西田幾多郎. 1950. *An Inquiry into the Good* 善の研究. Tokyo 東京: Iwanami Shoten 岩波書店.

Nishida, Kitarō 西田幾多郎. 1989. "The Logic of Place and the Religious World-view 場所的論理と宗教的世界観." In *Nishida Kitarō Tetsugaku Ronsyu III* 西田幾多郎哲学論集 III: 299–397. Tokyo 東京: Iwanami Shoten 岩波書店.

O'Callaghan, Jonathan. 2015. "Bill Gates Says We SHOULD Fear a Robot Uprising: Microsoft Co-founder Says He 'Agrees with Elon Musk' on Dangers of AI." *DailyMail* (online), 29 January 2015. https://www.dailymail.co.uk/sciencetech/article-2931375/. Accessed 18 November 2015.

Rose, Lacey. 2005. "Nanto City, Japan: Robotic Companions." *Forbes* (online), 6 August 2005. http://www.forbes.com/2005/06/08/cx_lr_0608japan.html. Accessed 18 November 2015.

Šabanović, Selma. 2014. "Inventing Japan's 'Robotics Culture': The Repeated Assembly of Science, Technology, and Culture in Social Robotics." *Robot Visions, Social Studies of Science* 44: 315–341.

Schodt, Frederik L. 1988. *Inside the Robot Kingdom: Japan, Mechatronics, and the Coming Robotopia*. Tokyo, New York: Kōdansha International.

Sharkey, Noel. 2007. "Robot wars are a reality: Armies want to give the power of life and death to machines without reason or conscience." *The Guardian*, 18 August 2007. http://www.theguardian.com/profile/noelsharkey. Accessed 18 November 2015.

Sullins, John P. 2007. "When Is a Robot a Moral Agent?" Presented at Workshop on Roboethics, ICRA'07: 2007 IEEE International Conference on Robotics and Automation, April 10–14, at Angelicum University, Rome, Italy (submitted paper to this workshop).

Suzuki, Yutaka, Lisa Galli, Ayaka Ikeda, Shoji Itakura, and Michiteru Kitazaki. 2015. "Measuring empathy for human and robot hand pain using electroencephalography." *Scientific Reports* 5, Article number: 15924 (2015). http://www.nature.com/articles/srep15924. Accessed 18 November 2015.

Takanishi, Atsuo. 2007. "Mottainai Thought and Social Acceptability of Robots in Japan" (slides for the talk about this subject). Presented at ICRA'07: 2007 IEEE International Conference on Robotics and Automation, April 10–14, at Angelicum University, Rome, Italy. www.roboethics.org/icra2007/contributions/.../Takanishi_icra07_ppt.pdf. Accessed 31 December 2012(this site was not available on 5 December, 2015).

Tomiyama, Hidehiko 富山英彦. 2004. "A Study of Anthropology by Tetsurō Watsuji – Focusing on Negative Movement of the Relativity 和辻哲郎の人間学的考察~「個と関係の否定モデル」を手がかりにして~." *Journal of Tokyo University of Information Sciences* 東京情報大学研究論集 7.2: 27–44.

Turkle, Sherry. 2007. "Authenticity in the Age of Digital Companions." *Interaction Studies* 8.3: 501–517.

Turkle, Sherry. 2011. *Alone Together – Why We Expect More from Technology and Less from Each Other*. New York: Basic Books.

Ura, Ddaisuke 浦大介, Iribe Masatsugu 入部正継, Osuka Koichi 大須賀公一, and Kinugasa Tetsuya 衣笠哲也. 2015. "Legged Walking Robot Design Applying a Behavior of Passive Dynamic Walking: Joint D.O.F Alignment Design Applying the Adaptive Function 受動的動歩行の性質を利用した脚歩行ロボットの一設計方法-適応的機能を使用した形状と関節自由度構成の設計-." *Transactions of the Society of Instrument and Control Engineers* 計測自動制御学会論文集 51.5: 329–335.

Veruggio, Gianmarco, and Fiorella Operto. 2006. "Roboethics: a Bottom-up Interdisciplinary Discourse in the Field of Applied Ethics in Robotics." *International Review of Information Ethics* 6: 2–8.

Wagner, Wieland. 2005. "Land der Roboter." *Der Spiegel* 6:136–138.

Watsuji, Testurō 和辻哲郎. 1937. *Ethics vol. one* 倫理学(一). Tokyo 東京: Iwanami Shoten 岩波書店.

Watsuji, Testurō 和辻哲郎. 1979. Climate: Thoughts on the Study of Human Beings 風土-人間学的考察. Tokyo 東京: Iwanami Shoten 岩波書店.

Winnicott, D.W. 1971. Playing and Reality. London: Tavistock Publications.

Yamaoka, Fumitaka, Takayuki Kanda, Hiroshi Ishiguro, and Norihiro Hagita. 2009. "Model of Proximity Control for Information-presenting Robots." *Journal of the Robotics Society of Japan* 27.2: 230–238.

Yuasa, Yasuo 湯浅泰雄. 1990. *Shintai-ron* 身体論. Tokyo 東京: Kōdansha 講談社. The original edition is: Yuasa Yasuo (1987). *The Body: Toward an Eastern Mind-Body Theory*, edited by Thomas P. Kasulis, translated by NAGATOMO Shigenori and Thomas P. Kasulis. New York: State University of New York Press.

Chapter 9
An Intuition-Based Approach to Sustainable ICT: Insights from Eco-Ethica

Per Fors and Mikael Laaksoharju

1 Introduction

Within fields interested in the relationship between sustainability and Information and Communications Technology (ICT), the fact that ICT can have both positive and negative effects on the environment is often emphasized. On the one hand, the production, use and disposal of ICT products are sources of environmental degradation. In the use phase, ICT products consume electricity that is not always produced from renewable energy sources. The production phase is not only highly energy demanding, but pollutes global and local environments. The disposal phase is problematic due to the difficulties of recycling ICT products, which has led to an illegal market of electrinic waste (E-waste), where worn out ICT products are exported to peripheral areas. On the other hand, many researchers and practitioners have in recent years started to focus on the potential of using ICT as Green Tech, that is, using ICT in order to improve the environment. Within areas such as agriculture, transportation and production of consumer goods, it has been calculated that ICT could potentially decrease the annual greenhouse gas emissions by 12 Gt CO_2e worldwide, and save 25 billion barrels of oil every year by 2030 (GeSI 2015).

The original version of this chapter was revised: MS Mincho font was updated throughout the book for Japanese and Chinese characters. The correction to this book is available at https://doi.org/10.1007/978-3-319-59027-1_13

Per Fors (✉)
Department of Engineering Sciences, Uppsala University, Uppsala, Sweden
e-mail: per.fors@angstrom.uu.se

Mikael Laaksoharju
Department of Information Technology, Uppsala University, Uppsala, Sweden
e-mail: mikael.laaksoharju@it.uu.se

Thomas Taro Lennerfors, Murata Kiyoshi (eds.), *Tetsugaku Companion to Japanese Ethics and Technology*, Tetsugaku Companions to Japanese Philosophy 1,
https://doi.org/10.1007/978-3-319-59027-1_9

In this chapter we draw on the Japanese philosopher Tomonobu Imamichi 今道友信 in order to further develop our understanding of the relationship between ICT and ecological sustainability. Our theoretical framework is inspired by his ethical framework Eco-ethica, presented as "a new virtue ethics by the times that lie ahead of us, a new ethics for a new dimension of human life (Imamichi 2009: *v*)." Eco-ethica was developed in order for ethics to catch up with the rapid development within the areas of science and technology during the second part of the twentieth century. According to Imamichi, a technology-mediated environment (TME) emerged alongside our natural environment when technologies started to communicate with each other without humans as intermediaries. This new environment can be described as a technological infrastructure – acting as a filter – where the relations between human and nature and in-between humans are altered. Imamichi argues that in the past, technologies were mainly used to facilitate existing human activities, which meant that traditional ethical theories were still valid. However, increasingly powerful and complex technologies seem to require new ethical reasoning. As we will argue in the following pages, this is particularly evident within the field of ICT and sustainability. Going forward, we will present carefully selected aspects of Eco-ethica and discuss how they could contribute to a more proactive and human-centered approach towards issues within the field.

We felt compelled to draw on this particular ethical framework for studying ICT and sustainability for several reasons. First, Japan is one of the world's leading nations in technological research and development. For several decades, Japan has been a hub for technological innovation, and the technological optimism has been stronger there than in most other developed countries (Odagiri and Goto 1996). Second, despite the widespread adoption of modern technology in the Japanese society, some practices are still performed with traditional tools in a conservative manner, in which the process of performing is considered equally important as the outcome. The most well-known examples include recreational activities like *shodō* 書道, *bonsai* 盆栽, traditional music, and martial arts like *kyūdō*弓道 and *naginata* 薙刀. Also within cooking and cleaning practices, traditional methods are used alongside hypermodern appliances. One notable historical detail along this line is that the Japanese abandoned the use of firearms for several hundred years, after first having been introduced to them by the Portuguese in 1543 (Nye 2006, see also the introductory chapter to this book). Third, Japanese philosophy is strongly influenced by the indigenous *Shintō*神道 and other related religions such as Chinese Confucianism (*jukyō* 儒教). These religious elements are contributing with a strong link to nature and the natural, and are often less anthropocentric than Western religions, even though the core of Confucianism is humanistic (Juergensmeyer 2005). For example, within *Shintō*, *Kami* 神 are worshipped as the energy which constitutes humans, animals and trees, but also non-living things such as rivers and rocks. In Confucianism, nature is seen as intrinsically valuable, thus not only a source of natural resources. Confucianism is "characterized by four key elements: an anthropocosmic rather than an anthropocentric perspective, an organic holism of the continuity of being, a dynamic vitalism of material force... and a comprehensive ethics embracing both humans and nature (Tucker 1998: 412)." Being able to see the role of the human in relation to natural processes is critical within this worldview, and this is strongly related to sustainability. The contrast between the strong technology optimism of recent decades in Japan on one hand and the philosophical roots emphasizing the delicate relationship between man and nature on the other enticed us to look deeper into Japanese philosophy and Eco-ethica in particular. Last, unlike many other (Eastern) ethical theories, Eco-ethica was developed in relation to and within the TME.

In this chapter, we ask how a sustainable approach to ICT could look like within the TME. We argue that, with an analytical starting point in Imamichi's reversal of the practical syllogism, the way to promote a sustainable development of ICT is to design technology in a way that allows users to better conceptualize the negative environmental impacts of various stages of the ICT life cycle, as these impacts are usually systematically obscured from our awareness, by for example outsourcing. Conventional approaches include persuasive technologies aiming to raise awareness or nudge (Brynjarsdóttir et al. 2012; Knowles et al. 2014; Thaler and Sunstein 2008), resonant interactions and worldmaking (Bendor 2018). While we are sympathetic towards many of these approaches, we suggest an intuition-based design approach that aims to invoke feelings of wastefulness and misuse if the product is used in an unsustainable manner. ICT's designed using this approach make environmental consequences perceptible and possible to fathom for the user, without her having to grasp their inner workings and life cycle effects, or force her into certain pre-determined actions by means of nudging or persuasion. Based on tangible feedback that caters to our natural pattern-matching, intuitive thinking processes (Kahneman 2003, 2011), it thus allows even "skilled animals" (Imamichi 2009), that is, users unaware of the inner workings of a particular piece of technology, to make ethical and sustainable decisions in relation to ICT. In turn, we believe that their choices may affect the offerings of producers.

2 Technology and Sustainability

Technology has always been perceived as a double-edged sword with both positive and negative impacts on our society. Efficient production in factories during the industrialization enabled the lower and middle class to purchase and use products that were previously only available for the upper classes. Factories employed thousands of people in and around cities, many of them previously unemployed. The urbanization that followed also benefitted cultural development, as cities became creative hubs for artists, musicians and poets. However, the industrialization also resulted in many negative social and environmental side effects. The coal-powered production plants covered large cities in thick smog, resulting in serious health issues. Toxic waste from factories polluted water supplies and nearby landmasses, and the monotonous and exhausting work activities of the labour force caused workers to face untimely deaths. As the positive effects of industrialization were too valuable for the capitalist classes to refrain from, the technological development had to continue despite the negative social and environmental effects. Moreover, as technology and society advanced, many side effects could be mitigated by regulation and cleaner and more efficient energy sources and technology.

Despite the fact that technological development has always met resistance because of its negative side effects, the ICT sector has historically been spared much of this critique (Lennerfors et al. 2015). While the computerization was opposed by some in the late twentieth century, due to fear of workers being replaced by computers and robots, and of radiation from CRT monitors, people quickly accepted ICT as a natural part of their work environment (Boivie 2007).

Some concern was also expressed by the environmental movement because of the negative impact of ICT products on the external environment (Nordin 2010). In the production phase, raw materials such as oil, metals such as copper, and Rare Earth Elements (REEs) are required. ICT equipment accounts for about seven percent of all plastic consumer waste (Cramer 2012). When extracting copper, 600 tons of waste is generated for every ton of copper extracted (Gulbrandsen 2012). Many components are produced out of REEs such as tantalum, cobalt and beryllium which are not only rare, but also hazardous and energy demanding to extract (Cramer 2012).

The question of CO_2 emissions due to the high energy demand of ICTs has increased its significance recently. In Sweden, the total carbon footprint of ICT is estimated to about two percent of the total carbon footprint (Malmodin and Lundén 2016), in par with the airline industry. According to many researchers, the carbon footprint of ICT will continue to rise worldwide, especially in peripheral areas.

There are also issues related to the afterlife of ICT products. While a small portion of obsolete ICT products is being properly recycled, most devices end up in landfills or are illegally shipped as second-hand goods to rural areas of for example China, India, Nigeria and Pakistan (Umair and Anderberg no date). In 2008, about 275.000 tons of electronic waste was illegally exported (Kostigen 2008). Informal recycling involves highly toxic processes resulting in human exposure to mercury, cadmium and arsenic, and is often carried out in rural areas by poor people, including children. While this pollutes local environments in developing countries, most users of ICT will never see nor experience these effects first-handedly.

However, it has recently been suggested by researchers, politicians and practitioners that implementing more ICT-supported practices could have a net positive impact on the environment. Reports state that ICT has the potential to reduce the global greenhouse gas (GHG) emissions by up to 20 percent globally, or by 12 Gt CO_2e, especially within areas such as agriculture, transportation and production (GeSI 2015). Because ICT can have both positive and negative effects on sustainability, some of which are hard to evaluate and measure, the topic is considered highly complex. In previous articles, we have however argued that these impacts are unequally distributed throughout the world-system (Lennerfors et al. 2015; Fuchs and Horak 2008; Hornborg 2001). While the positive (social and economic) effects of ICT are enjoyed where ICT is used, the negative (social and environmental) consequences have been systematically moved away from the main users through exportation of environmental disorder (Hornborg 2001). Production facilities first moved out of cities to peripheral areas of core countries, then out of core countries to low-wage, peripheral countries where ICT is used to a much lesser extent. Moreover, the production of the electricity used in ICT devices has also been moved away from the user physically. The same thing can be said about the illegally exported E-waste. This means that neither the production, nor the energy generation, the extraction of raw materials, and the disposal of ICT equipment are directly visible for most ICT users. The environmental footprint of ICT has been made invisible, hidden away like landfills, "out of sight, out of mind (Bull 2015: 20)."

We argue that as most consumers of ICTs are not experiencing these side effects first-handedly, they are struggling to take them into consideration when consuming and using ICT. On the contrary, positive effects provided by ICT, such as entertainment value and increased efficiency, are enjoyed by the user and, consequently, easy to take into account. The decision to use ICT for a variety of purposes thus becomes easy to justify.

In the following section we will present three concepts from Eco-ethica addressing how technology and ICT in particular has radically changed the conditions on three levels: the individual, the organizational and the societal. By dividing the issue into these levels and discussing them in relation to each other, we subsequently propose alternative forms of sustainable ICT design, inspired by intuition rather than information or first-hand experience.

3 Eco-Ethica

Turning to Japanese philosophy in studies on ICT might seem odd for two Scandinavian researchers. Many Western thinkers have historically claimed that philosophy is characteristic only to Western cultures. Martin Heidegger, for example, has explicitly expressed that only the Greek and German languages are appropriate for philosophizing (Farías et al. 1991). Although this might not be the general view, contemporary Eastern philosophy is rarely translated to English or any other Western language.

Tomonobu Imamichi (1922–2012) was a prominent Japanese philosopher focusing mainly on Chinese philosophy. Between 1953 and 1958, he was lecturing at universities in Tokyo, Paris, Monaco, and Würzberg. He received his first professorship in 1958 at the University of Kyushu. He was also vice president for the International Society of Aesthetics between 1974 and 1994 and was assigned chairman for Centrum Internationale pour l'Etude et Philosophy in 1997. He devoted the last part of his career to spread and develop his concept of Eco-ethica, which he developed during the 1960s and 1970s.

In the following sections, we will present three concepts from Eco-ethica, namely the *technology-mediated environment*, the *skilled animal,* and *the reversal of the practical syllogism.* Although they are presented separately, they are all interconnected, as both the reversed practical syllogism and the skilled animal are effects of and require the TME on an organizational level and individual level respectively. As organizations consist of individuals, the skilled animal approach affects organizational decision-making, and organizational decision-making affects the individual and how she behaves in relation to technology developed by organizations. By using these concepts from Eco-ethica to understand how ICT and sustainability are related, we subsequently construct our own approach to sustainable ICT design based on intuition, by drawing on our previous research on ICT systems design (Laaksoharju 2014), influenced by the philosophy of Heidegger (1977) and the two systems model of cognition (Kahneman 2003, 2011).

3.1 The Technology-Mediated Environment

According to Imamichi (2009: *x*), the TME "has become established alongside nature as a new human environment, evoking expectations of an ethics that differs from the behavioral norms established when our environment consisted solely of nature." The TME has thus not always existed but emerged once technological appliances became interconnected. Imamichi argues that the TME has altered our relationship with nature and other human beings, and that the emergence of this environment calls for new ethics for a number of reasons. First, it has radically changed the way we have to approach person-to-person ethics, as illustrated by the following example:

> In the everyday act of speaking, for instance, our actual neighbors are in close physical proximity to us, but with the technology-mediated environment people in far-flung corners of the globe become our neighbors via the telephone. So inadvertently dialing a wrong number or getting the time zone wrong means that you end up waking a complete stranger in the middle of the night, far off in some foreign country. (Ibid: *x*)

Because of the fact that this particular technological system, in this case the telecommunication network, enables us to contact any person at any time just by lifting the handset, the majority of the inhabitants of the world could potentially be as close to us as our physical neighbors, "broadening... ethics beyond a specific minority of visible neighbors to an unspecified majority of invisible and unknown neighbors (ibid.: *vii*)." It can be described as a technology that decreases temporal and social distance while the spatial distance remains unchanged (Trope and Liberman 2010). Even though ethics had to take into account people other than our physical neighbors also before the telecommunication network was invented, technological innovations such as this increase the number of people that must be taken into consideration in ethical reasoning and moral decisions.

Second, the TME has altered our relationship with nature. In the past, our society was dependent on natural aspects and phenomena when making decisions on where we could live, when we were able to travel, and what we could eat. Thanks to technological innovations, this is no longer the case. The example of the ancient pit dwellings of Tokoro 常呂 from the same book illustrates how our habitat is no longer dependent on running water, the climate, and other natural aspects:

> All of these pit dwellings – said to number two or three thousand in total – are located on low hills... The hills receive good sunlight from the south, and to the north there are woods, which act as a natural wind barrier. Nearby is a river known as the Tokoro River... [and] the water needed for daily life could be obtained simply by digging downwards... Today, it's ideal if these... natural conditions are met, but as long as air conditioning and gas and water facilities are available, you can live without a river nearby, and in a house that faces north and does not receive good sunlight. (Imamichi 2009: 17)

As the above two examples illustrate, the TME has changed the way we interact with nature and other human beings, which has complicated things in terms of ethics. Person-to-person ethics is no longer local but global, and considering how many of us behave online, it is obvious that many of us still have not adjusted to this revolutionary new way of interacting with our neighbors. Even more relevant for this chapter is the observation that the TME has reduced our need to adapt to nature and its forces, as we now use technology to appropriate our natural habitat. We do not have to consider the natural forces, water and sun, and other aspects of nature in order to live. This suggests that nature has been tamed by the means of the TME.

Third, the TME has changed how we interact with technology. We have been more or less forced to incorporate new types of behaviors and practices in order to make ourselves understood by the machine. As suggested by Imamichi, this tends to give the impression that "human beings are becoming like cogs in a machine (ibid.: 59)" (compare Charlie Chaplin in *Modern Times*). To take a mundane example, we are encouraged to use *like* functions to signal approval in social media. Thus, service providers can understand what we are interested in and target us with relevant advertisement. What is worse, according to Imamichi, is that such "wordless interaction with mechanical devices basically involves us turning into things that associate with other things via codes (ibid.: 59)." Also, "since signals demand an accurate and prompt response, they deprive us of the habit of training ourselves to make decisions (ibid.: 60)." Imamichi's drastic conclusion is that humans lose the quality that distinguishes us from other animals: "responding directly to stimuli means that we are no longer human (ibid.: 60)."

3.2 *Turning into Skilled Animals*

What does Imamichi mean by saying that we are no longer human in the TME? He thinks that because of the increased complexity of today's technological artifacts, most humans use technology like trained or skilled animals would have. That is, without reflecting on how these artifacts actually work or how using them affects our surroundings:

> When I enter the realm of machines and stand in front of what's popularly referred to as a 'black box', I feel as if I've understood it just by simply getting the hang of operating the black box, even though I have no inkling of its inner workings... apart from knowing how to operate it I have absolutely no understanding of how the setup works. Only those who produced the device have any idea... This is exactly the same as how animals get into trouble if they don't wait to eat when told to do so and how they start eating once given the okay, even though they have no understanding of the words at all... If we suppose that human beings are moving in a similar direction, then in a sense we are descending to the level of skilled animals in our new society on the technology-mediated environment. As long as we too pretend to understand despite not having a clue, comprehension degenerates into a matter of being able to operate something. (ibid.: 27f)

In a sense, understanding technology and its consequences has become redundant in the TME. To achieve the desired functionality of the machine it is sufficient that the user provides the correct input, often by pressing the right buttons on the interface, so that the technology can understand her. Regardless of what happens within the machine, a corresponding output will be delivered if the technology can manage to understand us. A push of a button could mean that the music from a CD starts playing, or an intercontinental ballistic missile is launched from a missile launch facility.

However, even though both the inner workings of technology and its indirect consequences of use are obscured from our direct perception, they exist. Its inner workings have merely been blackboxed, while the most negative effects of the ICT life cycle have been exported to peripheral areas of the world system (Lennerfors et al. 2015). This is related to the concept of commodity fetishism, which means that people only see the characteristics of the end product rather than the process in which the commodity itself has been produced (Hudson and Hudson 2003). In other words, the social and environmental implications of the product remain invisible to the typical user, making it practically impossible to assess and address such issues.

We argue that the commodification process could be seen as a result of the skilled animal approach to ICT and technology in general. Even though we agree on the fact that ICT is being commodified, we suggest that a more ethical approach to using and consuming ICT must be possible even within a capitalist system, despite a suspicion that this would be undesirable for most ICT producers and users. The skilled animal cannot fully understand ICT, and constantly assessing all parameters of the ICT life cycle would indeed further complicate decision processes. In an efficiency-oriented society, there is no justified need for the user to know how the technology was produced and under which circumstances, how much electricity the equipment consumes and whether the equipment will be recycled or not once it has stopped working. We relate this to Alvesson and Spicer's (2012: 1194) notion of functional stupidity, which refers to "an absence of reflexivity, a refusal to use intellectual capacities in other than myopic ways... not through intellectual deficits but through political expediency and the operation of power". While functional stupidity is a way to increase efficiency by consciously delimiting choice and avoid the need for careful reflection, the skilled animal instead represents an attitude toward technology in the TME. By restricting us to only see the relation between the input and the output, using or not using contemporary ICT becomes a non-issue for the skilled animal. If we would take everything into account, from the production to the disposal of the product, as presented in the previous section, using ICT would be much harder to justify.

Furthermore, as skilled animals we do not need to acquire expertise to accomplish remarkable things – we only need access to a computer with the appropriate software. If we borrow one of Imamichi's examples we can observe that a complete novice, aided by computational software, can perform many calculations faster and more accurately than a math professor (2008). If we focus exclusively on outcome of effort, it appears wasteful to work hard for a full life in order to develop expertise that is made redundant by machines. Thus, from the point-of-view of a skilled animal, technology does and should exist as a mean, and we only have to choose the purpose for which it should be used.

3.3 Reversal of Practical Syllogism – The Primacy of the Mean

Imamichi has garnered most international recognition for proposing a new form of practical syllogism to describe how decisions are made within modern organizations. What he observed was that a choice no longer necessarily followed the practical syllogism, originally formulated by Aristotle, namely that we choose between means to achieve desired ends. Instead, Imamichi argues, we are in a situation in which many choices are made based on a reversed logic: the means that we have at our disposal allow us to choose a variety of purposes to fulfill, of which we choose one.[1] Imamichi formalizes it in the following way (Imamichi 1998, 2009)[2]:

Major	We have the power *P*
Minor	*P* can realize *a*, *b*, *c*, ... as purpose
Conclusion	We choose the purpose *a* because it is the (noblest and) most economically effective

A concrete example may serve as an illustration:

Major	We have efficient computing power
Minor	Efficient computing power can realize faster computers, cheaper computers, and reduced energy consumption as purpose
Conclusion	We choose faster computers because it is the (noblest and) most economically effective

Through this reversal of the logic of practice – from fulfilling human goals to utilizing available means – Imamichi wants to focus attention on that the technological

[1] The practical syllogism in itself seems to have become something of a simulacrum, i.e., it has evolved over time into a meme that no longer necessarily resembles its original formulation. It is used as a description of the reasoning involved in deciding what to do in a concrete situation. One common interpretation of the practical syllogism is that the major premise is a universal assertion (or opinion), the minor premise a particular observation (or opinion), and the conclusion an action, i.e., a practical decision.

Major	Universal assertion
Minor	Particular observation
Conclusion	Action

Imamichi's framing of the syllogism is slightly different. He recognizes that we make a choice when we act and that choices are what determine character rather than opinions (Imamichi 1998, compare EN3.2.1111b 26-EN3.2.1112b 17 with EN 7.3.1147a23–b5). Note that the minor premise in this version includes the major premise in the more common interpretation, and that the particular observation is implicitly assumed.

Major	I want *A*
Minor	Means *p*, *q*, *r*, ... can realize *A*
Conclusion	I choose the mean *p* because it is the noblest and most expedient

In his conclusion, choosing a mean represents acting. Furthermore, Imamichi also interprets the major premise as an end, rather than as any universal opinion and the minor premise as a selection of available means rather than as any particular opinion.

[2] In the reversed formulation, means have been replaced with powers, which probably should be understood as generalized means, i.e., potentials that technological systems and artifacts afford.

society has strayed from what he sees as humankind's natural way of living, and instead works according to the logic of closed systems. Idiosyncratically he has named this tendency *nosism*, a term which implies a collective form of egoism. The two negative implications of nosism are that (1) individual members of an organization are neither able to control, nor take responsibility for, the actions of the organization, and that (2) organizations, as abstract entities, will strive to maximise their intended utility for their members rather than assuming the moral responsibility that individuals conform to, including responsibility for a larger context (Imamichi 2008; Murata and Orito 2008). For instance, when we allow a democratically elected government to make decisions for us, and its officials have to make decisions according to what gets them re-elected, we have a situation of the reversed practical syllogism. The government has several means at their disposal that they can use to fulfill diverse purposes, from which they choose the purpose that gets them re-elected, which is not necessarily the one that is best for the citizens, nor for the planet.

Considering that Imamichi's arguments point to the regrettable consequence that humans have lost their role as autonomous ends in themselves, it becomes interesting to discuss implications of the reversal of logic. First, the nosist collective *we* in the reversed formulation implies that individuals are not in power of important decisions, which could be interpreted as a deterministic perspective. Second, it could be argued that the construal of many groups in the modern society are technological, in the sense that they are artificially constructed to maximize certain values, for instance profit for business organizations and justice for legal organizations, under the constraint of bureaucratic regulations. This makes it relevant to briefly evaluate how Imamichi's ideas relate to the school of thought that considers technology to be deterministic, that is, following its own evolutionary path regardless of human interests and, consequently, is driving societal change.

First, however, we would like to add a small caveat regarding the soon century-long debate over technology determinism; that it may have been held mostly against strawmen. Opponents against such a perspective accuse its advocates for attributing agency to technology, while most of the alleged advocates rather observe that technological development, also when it is undesirable, is driven through interactions in, and between civilizations. A stark example of this is the nuclear arms race during the Cold War. Very few seem to embrace the actual idea that technology has its own agency. Even the vocal opponent David Nye admits a qualification to the idea that people can fully govern the technological development when he in passing mentions that "a society or a group that is able to *act without outside interference* can abolish a powerful technology (2006: 18, emphasis added)." Outside interference may force a society, or individual, to adopt a technology that is only desirable in relation to the outside interference. This appears to be a fundamental position in Imamichi's thinking, which is reflected in the reversal of the practical syllogism through nosism.

In Imamichi's framing, the question is not whether technology has agency in itself but whether the technological infrastructure that we have created has taken the role of an *outside interference* so that it, instead of only serving us, has come to also influence our decisions by providing us with new tempting means. In other words whether we as individual human beings are pointing out the direction for development, or, if we are guided by the TME in the sense that we cannot resist making use of it. When the question is framed like this, which we understand to be Imamichi's position,[3] the pertinent issue becomes to understand what the consequences are of technological frameworks that are not piloted by humans but instead emerge from interactions between individuals, organizations, and technology.

So what does Imamichi mean when he claims that we choose the purpose based on available means, especially in the light of his definition of *nosism* as collective egoism? In the reversal of the practical syllogism, *we* do not refer to a group of individuals but to the emergent, abstract entity that serves its members's (the nos's) desire to fulfill a particular purpose. This entity interacts, collaborates, and competes with other similarly constructed entities. This is an uncontroversial way to describe modern organizations, but what Imamichi seems to argue is that such emergent entities come to construct their own purposes based on what is the most effective usage of their powers. This can indeed be observed in many parts of contemporary society, in the form of decisions that are rational for the entity, but not necessarily for its members or the larger context. Examples of this include the governmental end-of-year spending to avoid budget cuts (Liebman and Mahoney 2013), marketing,[4] and the shifted purpose of the patent system.[5] Such phenomena are widely considered as non-optimal, yet they are effects of decisions that appear perfectly rational for the organizations involved in the corresponding system.

A domain in which we can observe a strong tendency of means supplanting purposes is politics. On the most general level, regarding taxes, we can observe that the salient question is usually not why a government collects tax, but how the tax money can be put to best use. The power, in this case in the form of economical resources, is taken for granted so the problem transforms into deciding what purpose it should fulfill. Furthermore, in Sweden, a form of tax relief for purchasing household assistance was introduced in 2007. The purpose of this subsidy was primarily to reduce the amount of undeclared work and stimulate entrepreneurship within the area (Skatteverket 2011). Initially the tax relief measure was using the power of taxation to fulfill a specific purpose, but over time it has turned into an independent power that the government can use to influence its citizens' behaviors. The Swedish gov-

[3] "[Our civilization] is a technological space that consists of a technology-mediated environment supported by the dual structure of the *modern nation* on the one hand and *globalization* as a negation of this on the other hand (Imamichi 2009, 3, emphasis in original)."

[4] Within marketing, a product is given but not necessarily the purpose of the product. The power (the manufacturing of the product) is used to target various purposes, based on what seems to be economically most beneficial. Meal replacements, for instance, can be claimed to save time, improve health, reduce environmental impact, etc.

[5] The original purpose of the patent system was to avoid knowledge getting lost when an innovator passed away, by granting him or her an exclusive right to use a technique in exchange for revealing it for posterity. The power, i.e., the patent, is the same but the purpose is now replaced by other purposes like preventing competition, or demanding a share of someone elses production by claiming patent infringement.

ernment has, consequently, put forth proposals to use the same tax relief system to reduce toxic waste by subsidizing reparation of household appliances, and increase employment by professionalizing maintenance of ICT equipment in households and gardening (Finansdepartementet 2015, 2016; Ekonomistyrningsverket 2016). The available mean affords selection of purposes to fulfill. A contrasting interpretation of this phenomenon is to regard it as a function creep, that is, a power initially devised for a specific purpose that is utilized for another purpose as well. In this perspective, the purpose retains precedence over the mean. Which of the two different logics at play is impossible to judge from the observation of a phenomenon, as it is dependent on the particular decision process. In this case it is not unlikely that the decision process followed the reversed logic, that is, that the tax relief was a powerful political instrument, waiting to be used for new purposes, rather than that there was a pressing need to create jobs within for example ICT maintenance or gardening.

Measures like in the above example may not be unjustified, and it is not the intention here to criticize specific applications of the reversed logic. However, if the purpose is for instance to reduce waste, other means may be more effective than tax subsidies for reparation of household appliances, but when following the reversed logic, such means simply do not surface.

It seems appropriate to briefly pause here to consider whether the reversal of the practical syllogism is just a different way of framing common economic rationality in neoclassical economic theory, which, at least in theory, is an effective way to allocate scarce resources in an optimal way. The nosism described by Imamichi corresponds well with the free-market advocates' interpretation of a corporation's moral responsibility; to produce profit for its shareholders (see Friedman 2007), and what is called resources in standard economic terminology is easy to relate to Imamichi's definition of powers. However, a problem if we were to equate the two, which is especially pressing when discussing sustainability, is that capitalism does not have any built-in mechanism for determining the limit of growth, that is, when any possible reallocation of resources still leads to unsustainable expenditure of natural resources (Daly 1990). For this problem Imamichi's syllogism may provide an avenue to a solution by generalizing the purposes for organizations also in a capitalistic system. It does not change the premise that organizations exist in order to benefit their members, but benefits do not necessarily imply or entail economic revenue. According to the reversed practical syllogism, organisations will choose the purposes that are economically most effective, which ultimately means those that correspond to the preferences of customers. Although Imamichi translates Aristotle's *most expedient* to *most economically beneficial* (Imamichi 2009), we argue that this simplification cannot always be assumed. If customers were given the insights necessary to make sustainable choices rather than purely economical, it would inevitably affect the value system of business organizations as well.

4 A Sustainable Approach to ICT in the Technology-Mediated Environment

Historically, we have been able to intuitively tell the effects on the environment and on the health of people when a particular technological artifact is used. Early factories were dangerous places filled with machinery, and accidents happened frequently during the industrialization. Moreover, people could see that what was coming out of the chimneys and wastewater pipes polluted air and water and made them sick. The fact that the negative consequences of technological development in today's society have been systematically obscured has implications for individual and organizational decision-making. A skilled animal cannot possibly be expected to assess all environmental consequences when making ICT related decisions. She will probably only take the most apparent parameters into consideration, such as the cost, performance and aesthetics of the product when consuming ICT. As she is not directly affected by the environmental consequences of the product, even receiving accurate information about for example electricity and resource consumption is probably not enough to make an ethical purchase decision. Thus, it might not be sufficient to simply understand the *inner workings* of technology in order to live a sustainable life (Imamichi 2009).

4.1 Teaching Skilled Animals New Tricks and Beyond

While we argue that a skilled animal cannot be expected to accurately assess the consequences of ICT on sustainability, we can see some potential in solutions utilizing this approach in the design phase. In the same manner as we teach dogs to sit or eat only when told to, technology users can also be taught new tricks by means of rewards and punishment. If we see that a particular behavior will benefit us in some way, we can adopt it. In ICT, this kind of learning can be achieved by the means of persuasion, for example through gamification and nudging (Deterding et al. 2011; Thaler and Sunstein 2008), which means that the technology itself suggests sustainable options by either making them more fun or by making them more easily accessible than unsustainable options. However, we have seen that persuasion such as gamification has severe limitations when it comes to dealing with sustainability (Brynjarsdóttir et al. 2012). While gamified applications can be effective in order to reach short-term goals, they often fall short on providing intrinsic motivation, essential for long-term commitment (Fors 2016; Fors and Lennerfors 2016). Although we have provided new, fruitful alternatives to gamification for sustainable purposes elsewhere (ibid.), we feel compelled to investigate alternative vistas for sustainable ICT design.

If we follow the distinction made by Imamichi between the human and other animals, persuasive technology – based on for example gamification and nudging – does not necessarily make us human, since we are still cluelessly responding to signals and other stimuli. To resolve this issue, we could perhaps try to bring our natural environment back into the TME by designing technology in ways that allow humans to formulate new intuitions about the functionality of ICT products. The natural environment affords predictability for the human – at least most of the time – because the human, through evolution, has developed purposeful senses and instincts to intuitively respond appropriately to recurring natural phenomena. The natural world is sufficiently transparent to the human, and it affords her successful residence. This transparency is arguably not complete. The human can, for instance, not observe molecules and atoms directly. Yet, evolution has optimized useful abstractions in the form of our senses, allowing us to smell, taste, see, and feel important properties even on the molecular level. Also when nature is deceitful, as in mimicry, the human can often use a combination of senses to make proper judgments, and if not, abstinence has always been a safe alternative.

Today, we rarely interact with nature, or other human beings even, directly, but indirectly mediated by technology. In the artificial TME, human senses and instincts are rarely utilized to their full extent, and are often maladaptive. Taste and smell are almost functionally redundant, as all food items are now labeled with expiration dates, and have instead come to serve largely hedonic functions. Vision and hearing are preoccupied by mediated attention wars. The human instinct to consume energy results in obesity in the "developed" world of excess, and the instinct to reproduce causes overpopulation. The human predilection for observing patterns lead to cargo cult thinking and the predisposition for inductive thinking leads to erroneous mental models of observed phenomena. Furthermore, the human inability to attend to slow processes, large-scale phenomena, and complexity, deprives us of proper understanding of the effects of global technological systems. We are left to hypothesizing about what impact the artificial world has on the natural, by mapping measurements of the state of the environment to a scale that is intelligible for us, and even then, the impact on our thinking can only be on an analytical level. The environmental consequences of using ICT do not affect us directly, neither can we feel or see these consequences using our senses, which becomes particularly evident in global environmental issues, especially global warming.

4.2 Unveiling the Environmental Consequences of ICT

A radical idea for visualizing the consequences of our choices on the natural world would be to expose ourselves to them directly: to not displace energy and food production to uninhabited areas, to not purify drinking water from pollution, and even to place the exhaust pipe of the car inside the cabin (see Schwarzenegger 2015). This seems absurd, of course, but it would definitely change our perception of pollution and exploitation of natural resources. In order to develop our intuition-based approach to sustainable ICT design and illustrate our thinking, we would like to present another sustainability-related case briefly, namely that of food waste.

Due to the abundance of food articles available at supermarkets, not everything is actually sold and consumed, but used to "display a feeling of overflowing richness and choice (Say No to Food Waste 2016)."[6] It has been estimated that about one third of all food produced worldwide is thrown away. Because of the purchasing power of most supermarkets, they can afford this behavior and still make a profit. This has resulted in the practice of dumpster diving, that is, "stealing" food from garbage containers outside supermarkets. Most supermarkets answered back by putting padlocks on their containers, denying dumpster divers access and transparency of their wasteful practices. However, as this was highlighted by the media, more people started protesting against this wasteful behavior, resulting not only in law changes in France, where throwing away edible food products is now illegal, but also in how supermarkets handle food waste. For example, Rema 1000 in Denmark replaced their buy-one-get-one-free and other quantity-based discounts for general discounts in all of their stores (Stop Spild af Mad 2016)[7] and Coop (2017) in Sweden launched their campaign *Weird vegetables* where odd-shaped vegetables were sold with 25 percent discount instead of being thrown away.[8] As a response to this social movement, supermarkets themselves thus further highlight the problem and present relevant solutions.

From a sustainability point of view, this is a positive example of the effect of the reversed practical syllogism. The power that the stores have at their disposal is a marketing and sales infrastructure, which is being used to fulfill the purpose of (among other things) reducing waste. Note that these are responses to consumers becoming aware of the prior extent of waste, and that it is not necessarily the optimal way, from a global point of view, of handling sustainability related problems. However, for the stores it is considered to be the most economically effective solution to an emerged problem regarding public appearance.

Compared to the case of food waste and dumpster diving, the complexity of ICT is one important bottleneck. We do not understand the inner workings of most technological devices that we confidently operate daily, and we are not yet pushing the providers of ICT to become more transparent. Thus, we have yet to see similar initiatives within the ICT sector. In relation to the food industry, we are still in the padlock stage of hiding problems, with few exceptions (for example Fairphone). One example is the planned obsolescence of smartphones. Smartphone providers constantly push new updates, slowing down old phones to the point where they become impossible to use properly, and if we opt not to update our operating systems, newer applications will not be supported. Regardless of how we address the issue of slow systems, we will eventually be required to upgrade our hardware (Lennerfors 2014). Although this is wasting the resources of the earth, we take as given and rarely question that smartphones will only last for a couple of years and

[6] https://web.archive.org/web/20180516135911/https://saynotofoodwaste.org/what-can-you-do/supermarkets/

[7] https://web.archive.org/web/20180516135640/http://www.stopspildafmad.dk/

[8] https://web.archive.org/web/20160506172429/https://www.coop.se/vart%2D%2Dansvar/knasiga-gronsaker/

that the hardware cannot be upgraded. Sustainability initiatives within the ICT sector, such as Green IT and Sustainable ICT, arguably overestimate the potential of ICT based solutions, which require even more ICT hardware to be implemented, while disregarding its many environmental and social problems.

5 An Intuition-based Approach to Sustainable ICT Design

Compared to making technology users experience consequences of their choices on their surroundings first handedly (for example by placing the exhaust pipe inside of the car cabin), a less radical way could be to enable them to develop intuitions about how their lifestyles impact the planet. This could be achieved by designing technological artifacts so that their essential attributes become intuitively intelligible (Laaksoharju 2014). Rather than obscuring the inner workings and the consequences of the development and use of technological artifacts, one could imagine a development where designers instead increase transparency by not obscuring, or even by supplying, perceptible cues about, for instance, electricity and resource consumption. Perceptible cues make it possible for a skilled animal to utilize its innate senses for developing new intuitions about technology's environmental impact. Consider for instance the feeling of throwing away food, seeing a running water tap or the lights left on in an empty room, or when recyclable bottles are disposed as combustible waste. Even though you are totally unaware of how electricity is generated, how water is purified and distributed, or the environmental impact of the food or recycling industry, seeing this wasteful behavior feels wrong. Such tangible feedback caters to our pattern-matching, intuitive thinking processes. In Japan, wasteful behaviour is often referred to as *mottainai* 勿体無い, which roughly translates to *What a waste!* The term has been used by the Japanese environmental movement since at least 2002. However, using ICT is currently not as strongly connected to these intuitive feelings of *mottainai* as the above-mentioned examples, even though using and consuming ICT certainly has a greater impact on the environment than for example running water taps.

As an example of ICT usage that we can develop intuitions about, consider the cooling fans on older computers, which accidentally supply information about energy consumption. If a piece of software consumes large amounts of processing power, the user will be made aware of it by the increased noise from the fan. This kind of feedback has, however, been perceived as undesirable so great development efforts have been invested in removing it. In contrast, the Power-Aware Cord[9] (The PAC) (Gustafsson and Gyllenswärd 2005) is a re-design of a standard electrical power strip to display the amount of energy passing through it by electroluminescent wires molded into the cord. When overloaded, it starts flashing. While this design is arguably intuition-based, they even make the comparison to a running water tap, development of similar products can benefit from the theoretical founda-

[9] https://poweraware.com/

tion presented in this chapter. For example, the brightness can be adjusted so that it does not disturb the user, which goes against our approach. Curb[10] is another innovation for increasing awareness about electricity consumption. A device, which is installed in the breaker box of a home, supplies detailed feedback about electricity usage and can be used to locate energy vampires. However, it requires you to actively request information through a companion app and the information, visualized by graphs and numbers, is intended to be attentively consumed. One could imagine a solution based on an intuition-based approach, which, not unlike The PAC, would provide information about electricity consumption whether you ask for it or not. If the breakers would glow according to the current passing through them, a mundane breaker box could be re-designed into an aesthetic object of information.

Most campaigns to promote a sustainable lifestyle aim, much like Curb, at increasing awareness, which means that they are targeting the subject's reflective, analytical System Two (Kahneman 2003: 2011).[11] While appreciating the value of these campaigns, we argue that it is not reasonable to convince the majority of users to use System Two thinking, but instead suggest targeting System One thinking. This is also done, often by persuasive design. However, persuasive design for sustainability purposes has been criticized (Brynjarsdóttir et al. 2012). We suggest that System One thinking can instead be targeted for developing constructive intuitions about the functionality and consequences of ICT in everyday decision-making. Thus, there would be no need to surpass the tendency to act like skilled animals in order to become sustainable subjects in relation to ICT, as the feeling of wastefulness would come naturally when equipment is used in an unsustainable manner. Developers would still be required to use System Two thinking when developing software and hardware for ICT products, not in the process of communicating normative conclusions, but in the challenge to reveal as much as possible of the environmental impact through the design of the product.

While Imamichi (2009) argues that simply responding directly to stimuli would mean that humans are reduced to skilled animals, here, we present the opportunities in such an approach to sustainable ICT design. Constructive intuitions of environmental consequences, perceived in a similar fashion as we perceive the natural world, will likely allow us to spend less cognitive resources on finding out the environmental impact of everyday ICT use, and more on conscious decisions on how and why we use ICT to address our needs. In the long run, bringing back *the natural* into the TME could ironically help us act less like skilled animals and more like humans in relation to ICT, including taking responsibility for sustainability issues.

Moreover, these intuitions will not only impact the way we use technology, but indirectly impact decision making within companies, as shown above. By drawing on Imamichi's reversed practical syllogism, we can see that from a producer's point-

[10] http://energycurb.com/

[11] The two systems refer to two different ways in which the brain forms thoughts. System 1 is fast, emotional and often unconscious, while system two is slow, logical and calculating.

of-view, most decisions are not in the hand of one or a few individuals within the producer organization. Instead, the organization will strive to maximize its intended utility for its members, which for the most part implies maximizing financial profit (Imamichi 2008; Murata and Orito 2008). Although this is a logical consequence in the current economical paradigm, we argue that organizations could be incentivized to strive towards purposes other than purely financial, such as sustainability, without necessarily changing the mechanisms of the global economic system. By the logic of the reversed practical syllogism: if individuals are empowered to consider sustainable development as a criterion in their decision making, organizations would use their existing means in order to achieve this particular purpose in addition to already existing purposes.

References

Alvesson, Mats, and Andrew Spicer. 2012. "A stupidity-based theory of organizations." *Journal of Management Studies* 49: 1194–220.

Bendor, Roy. 2018. "Interaction Design for Sustainability Futures: Towards Worldmaking." Draft chapter for *Digital Technology and Sustainability: Acknowledging Paradox, Facing Conflict, and Embracing Disruption*, edited by Hazas, Mike, and Lisa Nathan. London & NY: Routledge.

Boivie, Per. E. 2007. *Global Standard: How Computer Displays Worldwide Got the TCO Logo*. Stockholm: Premiss Förlag.

Bull, Richard. 2015. "ICT as an enabler for sustainable development: reflections on opportunities and barriers." *Journal of Information, Communication and Ethics in Society* 13.1: 19–23.

Brynjarsdottir, Hrönn, Håkansson, Maria, Pierce, James, Baumer, Eric P. S., DiSalvo, Carl, and Phoebe Sengers. 2012. "Sustainably unpersuaded: How persuasion narrows our vision of sustainability". Paper presented at SIGCHI Conference on Human Factors in Computing Systems, May 5–10, Austin Convention Center, Texas, USA.

Coop. 2017. "Frågor och svar om knasiga grönsaker." https://web.archive.org/web/20160506172429/https://www.coop.se/vart%2D%2Dansvar/knasiga-gronsaker/. Accessed: 16 May 2018.

Cramer, Benjamin W. 2012. "Man's need or man's greed: The human rights ramifications of green ICTs." *Telematics and Informatics 29*: 337–347.

Daly, Herman E. 1990. "Sustainable Development: From Concept and Theory to Operational Principles." *Population and Development Review 16*: 25–43.

Deterding, Sebastian, Dixon, Dan, Khaled, Rilla, and Lennart Nacke. 2011. "From game design elements to gamefulness: defining gamification." Presented at *the 15th international academic MindTrek conference: Envisioning future media environments*, September 28–30, Tampere University of Technology, Finland.

Ekonomistyrningsverket. 2016. "Yttrande över promemorian Skattereduktion för reparation och underhåll av vitvaror." https://web.archive.org/web/20180516140416/https://www.esv.se/contentassets/7ac227a99d7748c1978c51628fbeca58/remissvar-34-288-2016.pdf. Accessed 16 May 2018.

Farías, Víctor, Margolis, Joseph, and Tom Rockmore. 1991. *Heidegger and Nazism*. Philadelphia: Temple University Press.

Finansdepartementet. 2015. "Kompletteringar av RUT-avdraget." https://web.archive.org/web/20180516140329/https://www.regeringen.se/contentassets/3fa064a709c64c95a4c2080af2339217/kompletteringar-av-rut-avdraget. Accessed 16 May 2018.

Finansdepartementet. 2016. "Skattereduktion för reparation och underhåll av vitvaror." https://web.archive.org/web/20180516140416/https://www.esv.se/contentassets/7ac227a99d7748c1978c51628fbeca58/remissvar-34-288-2016.pdf. Accessed 16 May 2018.

Friedman, Milton. 2007. "The Social Responsibility of Business Is to Increase Its Profits." In *Corporate Ethics and Corporate Governance*, edited by Zimmerli, Walter C., Holzinger, Richter, Klaus and Markus Holzinger, 173–178. Berlin, Heidelberg: Springer.

Fors, Per. 2016. "ICT and Japanese Philosophy: Gamification for Skilled Animals". Presented at the 9th International Conference on ICT, Society and Human Beings, 1–3 July, Pestana Casino Park Hotel, Funchal, Portugal.

Fors, Per and Thomas T. Lennerfors. 2016. "Gamification for Sustainability: Beyond the Aesthetico-Ludological Approach". In *The Business of Gamification*, edited by Dymek, Mikolaj and Peter Zackariasson. Routledge.

Fuchs, Christian and Eva Horak. 2008. "Africa and the digital divide." *Telematics and informatics*, 25: 99–116.

GeSI. 2015. "#SMARTer2030: ICTsolutions for 21st century challenges." https://web.archive.org/web/20180510100348/http://smarter2030.gesi.org/downloads/Full_report.pdf. Accessed 16 May 2018.

Gulbrandsen, Christoffer. 2012. Stealing Africa. Why Poverty? Motion Picture.

Gustafsson, Anton, and Magnus Gyllenswärd. 2005. "The power-aware cord: energy awareness through ambient information display". Presented at the CHI'05 on Human factors in computing systems, April 2–7, Portland, Oregon, USA.

Heidegger, Martin. 1977. "The Question Concerning Technology." In *The Question Concerning Technology and Other Essays*, edited by Heidegger, Martin. London and New York: Garland Publishing Inc.

Hornborg, Alf. 2001. *The power of the machine: Global inequalities of economy, technology, and environment.* Maryland: Rowman Altamira.

Hudson, Ian, and Mark Hudson. 2003. "Removing the veil? Commodity fetishism, fair trade, and the environment." *Organization & Environment*, 16: 413–430.

Imamichi, Tomonobu. 1998. "The Metaphysical Tradition of East Asia and its Actual Relevance." Filosoficky Casopis, 46: 403–408.

Imamichi, Tomonobu. 2008. "Technology and collective identity: issues of an eco-ethica." In *The Humanization of Technology and Chinese Culture, Chinese Philosophical Studies*, edited by Imamichi, Tomonobu, Miaoyang, Wang, and Liu Fangtong. Council for Research in Values and Philosophy.

Imamichi, Tomonobu. 2009. *An Introduction to Eco-ethica.* University press of America.

Juergensmeyer, Mark. 2005. *Religion in global civil society.* Oxford University Press.

Kahneman, Daniel. 2003. "Maps of bounded rationality: Psychology for behavioral economics." *American economic review*, 93: 1449–1475.

Kahneman, Daniel. 2011. *Thinking, fast and slow.* London: Macmillan.

Kostigen, Thomas. 2008. "The underbelly of globalization: our toxic wastes exported to developing countries". https://web.archive.org/web/20180516140530/http://archive.ban.org/library/Features/080925_the_underbelly_%20of_globalization.html. Accessed 16 May 2018.

Knowles, Bran, Blair, Lynne, Walker, Stuart, Coulton, Paul, Thomas, Lisa, and Louise Mullagh. 2014. "Patterns of persuasion for sustainability." Presented at the *2014 conference on Designing interactive systems*, June 21–25, Simon Fraiser University, Vancouver, Canada.

Laaksoharju, Mikael. 2014. "Designing for Autonomy." PhD diss., Uppsala University.

Lennerfors, Thomas T. Fors, Per, and Jolanda van Rooijen. 2015. "ICT and environmental sustainability in a changing society: The view of ecological World Systems Theory." *Information Technology & People*, 28: 758–74.

Lennerfors, Thomas Taro (2014). Sustainable and Fast ICT: Lessons from Dromology, *Journal of Information, Communication and Ethics in Society*, 12: 284–297.

Liebman, Jeffery B., and Neale Mahoney. 2013. "Do expiring budgets lead to wasteful year-end spending? Evidence from federal procurement. National Bureau of Economic Research. The National Bureau of Economic Research." https://web.archive.org/web/20180516140634/http://www.nber.org/papers/w19481. Accessed 16 May 2018.

Malmodin, Jens, and Dag Lundén. 2016. "The energy and carbon footprint of the ICT and E&M sector in Sweden 1990–2015 and beyond." Presented at the 4th International Conference on ICT for Sustainability, August 30–31, Amsterdam Business School, Amsterdam, Netherlands.

Murata, Kiyoshi, and Yohko Orito. 2008. "Three Challenges for Japanese ICT Professionalism." Presenter at ETHICOMP, September 24–26, University of Pavia, Mantua, Italy.

Nordin, Håkan. 2010. *Green IT: From Problem to Solution*. Stockholm: Håkan Nordin AB.

Nye, David E. 2006. *Technology Matters – Questions to Live With*. Massachusetts: The MIT Press.

Odagiri, Hiroyuki, and Akira Goto. 1996. *Technology and industrial development in Japan*. Canada: Oxford University Press.

Schwarzenegger, Arnold. 2015. "I don't give a **** if we agree about climate change" [Facebook note]. https://web.archive.org/web/20180516092239/https://www.facebook.com/notes/arnold-schwarzenegger/i-dont-give-a-if-we-agree-about-climate-change/10153855713574658/. Accessed 16 May 2018.

Skatteverket. 2011. "Om RUT och ROT och VITT och SVART Rapport 2011:1." https://web.archive.org/web/20180516140034/https://www.skatteverket.se/download/18.2b543913a42158acf800024807/1359707860719/rapport201101.pdf. Accessed 16 May 2018.

Say No to Food Waste. 2016. "Supermarkets". https://web.archive.org/web/20180516135911/https://saynotofoodwaste.org/what-can-you-do/supermarkets/. Accessed 16 May 2018.

Stop Spild af Mad (2016). I fokus: Stop Spild Af Mad mobiliserer hele værdikæden til at mindske madspild. https://web.archive.org/web/20180516135640/http://www.stopspildafmad.dk/. Accessed 16 May 2018.

Thaler, Richard. & Sunstein, Cass. 2008. "Nudge: Improving decisions about health, wealth, and happiness." London: Yale University Press.

Trope, Yaacov, and Nira Liberman. 2010. "Construal-level theory of psychological distance." *Psychological review*, 117: 440.

Tucker, Mary E. 1998. "Religious dimensions of Confucianism: Cosmology and cultivation." *Philosophy East and West*, 48: 5–45.

Umair, Shakila, and Stefan Anderberg. No date. "Ewaste imports and Informal Recycling in Pakistan". Royal Institute of Technology.

Chapter 10
Imamichi, Interdependence and Technology-Mediated Action – The Case of Border Governance

Elin Palm

1 Introduction

A global digitalization and informatization have, in a profound way, altered and redefined human interaction, for example in the form of trade, exchange and travel patterns. While the world, increasingly interconnected by modern means of transport and by Information and Communication Technology (ICT), often is depicted as a global village (see for instance Carter 2001), evoking a borderless community, this metaphor is misleading. Although globalization (among many other things) means increased interconnectedness in terms of ICT-based communication and trade at a supranational level, this interaction and the benefits stemming therefrom are still limited to the activities of a resourceful few (Huysmans 2006). At the same time as ICT and the Internet have enabled swift crossnational communication, created a global Information society, and in many ways shrunk the distance between people of the world, this form of technological advancement has also carried a novel twist of the classical divide between resourceful and poor nations and their members (Britz 2013). Only a small clique of the village members is fortunate enough to be able to move freely across international borders. National borders have been relaxed within the European Union (EU), allowing EU citizens to move freely between the member states. At the same time, the external borders of the Schengen region have been reinforced by advanced surveillance systems and sophisticated ICT systems support EU immigration governance with the aim to protect the union from

The original version of this chapter was revised: MS Mincho font was updated throughout the book for Japanese and Chinese characters. The correction to this book is available at https://doi.org/10.1007/978-3-319-59027-1_13

Elin Palm (✉)
Department of Culture and Communication, Linköping University, Linköping, Sweden
e-mail: elin.palm@liu.se

Thomas Taro Lennerfors, Murata Kiyoshi (eds.), *Tetsugaku Companion to Japanese Ethics and Technology*, Tetsugaku Companions to Japanese Philosophy 1,
https://doi.org/10.1007/978-3-319-59027-1_10

unauthorized entries (Dijstelbloem 2009; Dijstelbloem and Meijer 2011) as well as to, with a greater accuracy, differentiate between desirable and undesirable migrants (Ajana 2015). An increased national and regional protectionism expressed in strict admission/immigration criteria and strict border control leave many third country nationals, and in particular unskilled members of developing nations few options to enter the EU in a legal fashion. Individuals who seek to enter the EU in an unauthorized fashion are typically exposed to great risks. Following a 2014 report made by the International Organization of Migration (IOM), the many failed crossMediterranean journeys and ensuing casualties make Europe the most dangerous destination for irregular migrants in the world (Brian and Laczko 2014).

For these reasons, rather than representing the condition of the contemporary man, the global village metaphor is foremost an illustration of the situation of resourceful individuals from developed nations and the possibilities that modern means of communication entail for them. Certainly, the world is in *their* pocket. They find international networking and travel significantly easier, surrounded by less bureaucracy and cheaper, than ever before. Since the largest part of the global community is made up by underprivileged individuals who are subject to strict admission criteria in the first world, *"mobility is a privilege that is unevenly distributed among human beings* (Pécoud and de Guchteneire 2006)."

In this chapter, an attempt is made to investigate how the ethics of ICT can be understood in the case of surveillance-based migration governance by means of Japanese philosophy. More specifically, ethically significant implications of current ICT-based migration governance in the first world are critically discussed by means of the eco-ethics of IMAMICHI Tomonobu 今道友信 (Imamichi 2009). While several scholars have been concerned with the societal and ethical impact of novel technology and the need for critical reflection thereof, Imamichi offers a more holistic approach in which he articulates an interdependence man – environment necessary for the long-term stability of today's technology-mediated environment. This approach can be used to give a better understanding for the complexity of the ethical problems connected with surveillance-based border governance. The case is made that an increased surveillance of cross-border movements contributes to expose a most vulnerable group – irregular migrants – to unacceptable risks. Inspired by Imamichi's reasoning regarding the need for a proper understanding of the interdependence between man and nature as well as between man and man, and for an ethical reflection regarding the use of modern technology, it is argued that (i) ICT-based migration governance is an example of the manufacturing of "untimely deaths," that (ii) mass displacement is interlinked with environmental conditions, and that (iii) the first world, both triggering and stifling migration, have certain moral duties towards migrants and developing countries. Looking at the technical systems used to reinforce migration management, it is shown that ICT-based migration governance worsens the condition of irregular migrants. Drawing on Imamichi, it is stressed that alternative ways of using ICT compatible with must be sought to safeguard a more sustainable future.

In Sect. 2, IMAMICHI Tomonobu's understanding of the relation between man and nature as well as between man and man is presented together with his views on technology-mediated action and moral responsibility. His reasoning serves as a basis for a critical discussion on the ethics of ICT-based migration governance in Sect. 3. Section 4 starts out from Imamichi's emphasis on interdependence to investigate the causes behind international displacement and transnational migration and is followed by a proposal for an equitable responsibility sharing regarding transnational migration in Sect. 5. Section 6 concludes the discussion.

2 IMAMICHI Tomonobu and the Technology-Mediated Environment

In this section, some of the key components in IMAMICHI Tomonobu's reasoning on moral agency and responsibility are introduced as a basis for a critical discussion on ICT-based migration governance.

As long as ethics is framed exclusively as a matter of the good life of man, it cannot provide a basis for a new and sustainable world order (Imamichi 1984). Moral cosmopolitan attempts to develop a global ethics that respects all human beings equally well are insufficient to that end. In addition to concerns over the human condition, ethics must take into account the good of non-humans as well and include all life in the world. Concomitantly, a new type of responsibility is required. In order to promote a good society and a sustainable world order, attempts must be made to establish equitable societies within which the needs and interests of all parties – non-human as well as human - are duly respected. It is crucial that and a deeper recognition for the value of the ecological system is fostered. This is the key message in the eco-ethica proposed by Imamichi (Imamichi 2009). Eco-ethica is broader in scope than both environmental ethics and ecological ethics. By means of modern technology, man can alter the entire ecosystem. This potential motivates Imamichi to critically analyze the technology-mediated society of today and to discuss what the role of technology ought to be. His eco-ethica is the new ethics for the technological environment (technological conjuncture), informed by and with respect for nature.

Imamichi deliberately refrains from using "ethics" because the term is too individual-centered. During the past 50 years, ethical problems have, according to Imamichi, emerged that cannot be resolved by the traditional ethics that was developed in and for a society where individuals where living (more or less in harmony with and) on the conditions of nature (natural society). Modern technology enables man to master and exploit nature to an unprecedented extent. Certain technologies, such as nuclear technology, allow man to fundamentally alter the living conditions of all beings. This capacity acquired by means of advanced technologies and technological systems requires, according to Imamichi, a new ethical framework that articulates and recognizes the relation man - technology – environment. Similarly, quests for novel ethical frameworks, better suited to the conditions dictated by technology have emerged within the field of ICT-ethics as well, where some have argued that a new ethics is necessary to deal with the issues generated by information and communication technology (see for instance Gorniak 1996). Against such claims, others have argued that a closer look at the problems at hand reveals that they are novel twists on age-old problems rather than radically new issues. The values at stake are not of a radically different kind why existing ethical theories still can provide adequate action guidance (Johnson 1999). Imamichi's stance however, is that modern technology has transformed the conditions in such a profound way, not only for mankind but for all life on earth, and hence, that a radically different ethical approach is warranted.

A move towards a less human-centered approach is proposed. A corner stone in his reasoning is that man is interdependent on the environment and on non-sentient beings and that, from this interdependence, a certain form of moral responsibility follows. In light of the current ecological crisis, he argues that a different understanding for responsibility is necessary. In his view, the crisis is a consequence of a technological superiority. For the longest time, technological progress has been considered a necessity and technological development has been exempt from moral constraints. While decisions regarding nuclear energy, such as the management of nuclear power plants and nuclear waste may be taken within a nation, in case of an accident, several nations may be affected as seen in the Chernobyl and the Fukushima Daiichi "accidents." Since those who make the type of decisions that have a fundamental impact on the environment and our living conditions, are members of groups rather than single individuals, Imamichi proposes that responsibility should be connected to collective identity. Historically, ethics has focused individual identity, agency and responsibility. Causal responsibility or blameworthiness has been associated with discrete individuals and the source of moral responsibility has been located in the free will of individual moral agents. That is, while most ethical theories place moral constraints on single individuals and hold individual moral agents accountable for their actions, issues like collective identity, group morality and responsibility have been rather controversial and disputed. For reasons of coherence and fairness, it has been questioned whether the same criteria should hold for collective responsibility as for individual accountability where intentionality play a central role. Imamichi argues that the significance of collective identity must be better recognized. In his view, the power to by means of technology shape and alter living conditions in a fundamental way requires a new form of ethics that can hold influential decision makers morally responsible. In eco-ethica, Imamichi associates both causal responsibility and blameworthiness with *groups* and locates the source of moral responsibility in the collective actions taken by these groups understood as *collectives*. By tying responsibility to clearly defined groups such as corporate committees, boards of enterprises and public institutions, those that have the power to significantly alter our living conditions and the eco-system can be held responsible for their decisions. Such decisions are neither made by single persons, nor by mere aggregates of single individuals. It is also stressed that committee members act in a *collective capacity* within the *public domain*. In order to understand this claim, a certain contextualization is needed. In the Japanese society, Imamichi explains, persons have different roles within the private and the public realm respectively and expectations and duties associated with these roles may be conflicting, even incompatible. Decision makers in boards and institutions are acting in a certain capacity: as part of their collective identity in public life. An important task is to foster an ethical awareness among decision-makers that includes the interests of society as a whole and respects the ecosystem in such decision-making. Following the 2012 Independent Investigation Commission on the Fukushima Nuclear Accident, the earthquake and the tsunami on 11th March 2011 has devastating effects but the catastrophe should not be seen as a natural catastrophe. It was a catastrophe created by human beings – a catastrophe that could have been foreseen and should have been prevented. Effects could also have been alleviated by a more effective response. It is stated that: *What must be admitted – very painfully – is that this was a disaster "Made in Japan." Its fundamental causes are to be found in the ingrained conventions of Japanese culture: our reflexive obedience; our reluctance to question authority; our devotion to 'sticking with the program'; our groupism; and our insularity.*[1]

While technology can be used to promote better living conditions, in many practical cases, the positive potential is not fully utilized and the benefits are not evenly distributed. Technological artifacts and systems are also used in ways that do harm instead of good. Rather than drawing on available and encompassing knowledge and instead of using the capacity of technological advancement to build away significant risks and prevent people from dying, modern technologies and technological systems are often uncritically accepted. Not seldom, technologies are socially accepted even if they are saddled with risks and even if they contribute to casualties. For example, cars are generally accepted despite significant numbers of leathal traffic accidents every year. Greater demands can be raised on the use of technology however. A zero-tolerance vision has been formulated by the Swedish agency for road administration that aims at a traffic situation where no single person dies or is seriously injured in relation to traffic. "Vision Zero" – often criticized for being utopic – recognizes that although costly and difficult to ensure, risks and deaths could be avoided by informed strategies in the advanced society. It stresses shared responsibility for road safety. Instead of primarily focusing the role of road users, the responsibility of road system designers is also included.[2]

The need for and potential in critical questioning of technology is emphasized by Imamichi. *"Although the world in which we dwell should under the banner of science and technology and human rights be a happy one, the reality is different* (Imamichi 2009: 3)." Furthermore, Imamichi writes that *"We are exposed to the threats of seven "untimely deaths" – i.e. starvation, death from accidents, death in war, death from pollution, death through terrorism, suicide and murder. These existed in the past but today, atrocious termination of human life, which should be protected through technology and human rights, are far more common than in any previous century* (Imamichi 2009: 3)." Risks and deaths are considered untimely for the reason that they, if properly addressed, could be avoided. At least, many fatal outcomes could be avoided in the first world and in other parts of the world where first world agents are involved. Certainly, all risks can not be strategically built away

[1] The official report of the Fukushima Nuclear Accident Independent Investigation Commission (2012). Executive Summary. The National Diet of Japan NAIIC. https://www.nirs.org/wp-content/uploads/fukushima/naiic_report.pdf

[2] http://www.visionzeroinitiative.com/

but in many cases, technologies with known adverse effects could be banned, replaced or significantly modified and resources could be allocated (differently) in order to reduce casualties, distributing technologies evenly to prevent the loss of lives. That technology is allowed to carry negative consequences and that a resources/technology are unevenly distributed are in many practical cases the results of a certain priority setting and of cost calculations. Rather than reinforcing first world standards that forbid certain risks and risk exposure in certain production processes, many multinational corporations choose to place parts of product development in countries with less demanding legislation (for instance when it comes to pollution). Given the proliferation and power implicit in technology, attempts must be made humanize technology. Technology should be used in ways more respectful both to nature and to human rights and the respect for human rights need to be strengthened.

Even if many aspects of Imamichi's reasoning deserve further articulation and discussion, for instance regarding the foundation of agency, the role of intentionality in relation to collective responsibility and the need and capacity of the novel virtues,[3] it offers a starting-point for critical reflection regarding the use of surveillance-capable ICT-technology used to differentiate between members and nonmembers. Importantly, rather than down-playing the moral worth of human beings, Imamichi upgrades the value of non-human life and makes a strong case for an ethics that better respects and protects nonhuman beings. Imamichi's expanded morality is rather radical since most theories tend to regard everything non-human as mere raw material, not ends in themselves (Russel 1948: 514). His eco-ethica approach is useful as a prism to critically reflect on the ethical justifiability of emerging technologies, technology-mediated action and moral responsibility for such action. Even if Imamichi originally was concerned with health technology and an increasing capacity to influence humanity in most fundamental ways, the reasoning is applicable on other technologies as well. Here, his reasoning will serve as a starting point to discuss the impact of ICT-based border governance in a situation of mass displacement and an unprecedented level of transnational migration. What is considered important in his reasoning is first and foremost his holistic approach and an often overlooked interrelatedness. From a certain interconnectedness between human beings as well as between man and environment, moral responsibilities emerge. Drawing on Imamichi's reasoning on interdependence and responsibility, an attempt will be made to spell out the relation man-technology-environment and moral responsibility in the particular context of mass displacement and immigration and border governance.

[3] Many of the concerns raised in the long-standing debate on collective responsibility are not developed in Imamichis' reasoning, for instance the role of intentionality. It is not clear if groups must meet the same stringent conditions of moral responsibility that individuals do – intentionality being a key aspect.

3 Technology-Mediated Immigration and Border Control

Before discussing the moral significance of the relation man-technology-environment, the capacity and impact of ICT-mediated immigration and border control will be discussed. Inspired by Imamichi's call for critical reflection regarding technology, and by his emphasis on the unreasonableness of accepting (some) risks and adverse consequences as mere givens or as unavoidable necessities of technologies that overall promote desirable ends, in this section contemporary EU migration governance and border control will be described. Intended consequences and *de facto* outcomes of ICT-based EU migration governance will be mapped as a backdrop for a discussion on the ethical acceptability of such conduct.

For more than a decade, EU Member States have cooperated on border security issues through Council Regulation 2007/2004/EC[4] that established Frontex: the European Agency for the Management of Operational Cooperation at the External Borders of the Member State of the European Union. Frontex supplements national-level border control and law enforcement, with a special focus on the external borders. It also supports EU neighbour countries in border protection, as under the EU-Turkey agreement and the more recent "Malta declaration" specifying collaboration between Libya and the EU. Under such agreements, FRONTEX assist local border guards with intelligence and technological advancement.[5] Another pan-European agency recently established is the European Border Surveillance System (EUROSUR) intended to concert all border surveillance activities within the Schengen region states. Border management should utilize an integrated surveillance system. Following the communication of this agency, the main priority is to protect the borders of the Schengen regions and to prevent irregular migration. EUROSUR will:

- Contribute to the management of migration flows by reducing the number of irregular migrants entering the Schengen area undetected;
- Protect and save lives at the external borders by considerably diminishing the unacceptable death toll of migrants at sea;
- Increase the internal security of the European Union by preventing serious crime at the external borders of the Schengen area.

Integrated information and communication systems are considered essential to achieve the ends stipulated by EUROSUR. In particular, it is stressed that high-tech surveillance is necessary to prevent casualties among irregular migrants. Technologies used for border security in Europe include databases and data-mining software, biometrics, surveillance and detection technologies.[6] It is for instance required of all apprehended migrants to produce fingerprint samples to be stored in

---

[4] http://frontex.europa.eu/assets/About_Frontex/frontex_regulation_en.pdf

[5] http://www.coe.int/t/democracy/migration/Source/migration/ProtectingMigrantsECHR_ESCWeb.pdf

[6] http://register.consilium.europa.eu/doc/srv?l=EN&f=ST%2015010%202006%20INIT

the EURODAC database (Broeders 2007; Broeders and Engbersen 2007). Regarding control of land borders, the Schengen handbook recommends border guards to conduct surveillance and apprehension of illegal immigrants by means of night vision devices, thermal cameras as well as camera and sensor alarm systems. Maritime border surveillance, it is proposed, should be supported by patrol vessels, helicopters and fixed wing aircraft supplemented with e.g. radars and camera surveillance systems.[7] Not only is a wide array of technologies permitted for use but even mandated in order for border guards to carry out their duties. Many of these technologies collect, process and transfer person-specific and person-related data which can give raise to privacy issues (Palm 2014). Employed in a context where the aims are to prevent irregular and illegal entry and to apprehend trafficking, surveillance-capable technology can function as an instrument to discriminate and to obstruct fundamental human rights. Advanced radar systems and drones could support individuals in distress at sea by detecting and assisting them at an early stage but, the primary aim of the two agencies mentioned is that of preventing unauthorized entry. Territorial protection is given precedence over individual human rights to seek asylum and to personal security. Once detected, apprehended migrant vessels may, according to the scheme above be escorted to the assumed port of embarkation – not necessarily in Europe. In effect, migrants on board may be prevented from an individual asylum process. Such practice may also violate the principle of non-refoulement. Certainly, the clash with fundamental human rights is not a problem specific to the technology employed to monitor borders but with the EU migration regime and directives underlying border management. By focusing on instrumented border control, it comes clear how problematic the exclusionary practice under this regime is. The effectiveness of the technology exceeds traditional low-tech control and brings out the consequences of EU immigration and border directives more clearly. Networked border intelligence enables an unprecedented control of cross-border movements, close scrutiny of individual migrants and of territorial borders.[8] In addition to low-tech operations such as the construction of physical barriers in the form of walls along the Greek-Turkish and Hungarian-Serbian borders and "manual patrolling"[9] in regions bordering sending- and transit countries, surveillance run *surreptitiously* (Dijestelbloem 2009; Dijstelbloem and Meijer 2011). For instance, drones with infrared sensor equipment enable *automated* monitoring of vast land- and maritime areas. Automated systems extend the coverage and continuity of monitoring conduct (Palm 2014) and the movements of migrants can be followed from a far distance. In effect, unauthorized entry is made significantly more difficult. The EU's continuous reinforcement of its external borders places an

[7] http://frontex.europa.eu/assets/Publications/Research/Ethics_of_Border_Security_Report.pdf

[8] The Eurodac system comprises the Central Unit, initially operated by the Commission but now managed by the EU Agency for Large-Scale IT Systems, a national unit in each Member State, and the infrastructure for transmitting data between national units and the Central Unit.

[9] Cf. Operations Mare Nostrum and Triton. Cf. http://www.marina.difesa.it/EN/operations/Pagine/MareNostrum.aspx and http://europa.eu/rapid/press-release_MEMO-14-566_en.htm

already vulnerable group – individuals without access to authorized ways of travel – in a most precarious situation (Gündoğdu 2015). Nongovernmental organisations focusing irregular migrants report that restrictive immigration policies paired with heightened border surveillance and more systematic crackdowns on human smugglers/traffickers alter the routes of migrants and smugglers/traffickers rather than stifle irregular migration and irregular escort services (Hernández-Carretero 2009; Pinyol-Jiménez 2012; Brian and Laczko 2014). Strictly monitored borders have increased undocumented migrants' dependence on smugglers and traffickers and made transportation of unauthorized travellers a most profitable business. Assisted by human traffickers, irregular migrants become vulnerable to life-threatening risks and exploitation (Pinyol-Jiménez 2012; Brian and Laczko 2014). Atrocities, sexual and monetary exploitation and abuse are common features of the smuggling/trafficking of human beings (cf. Brian and Laczko 2014). In that sense, the EU's ambitions to increase territorial security have come at the cost of severe infringements of the personal security of third country nationals. Since 2011 and the turmoil following the Arab spring, thousands of individuals have drowned every year when undertaking unauthorized journeys across the Mediterranean Sea. Certainly, all of these casualties are not the direct result of border surveillance measures but such activities typically make trans Mediterranean journeys riskier instead of safer although the safety of migrants could be enhanced if search and rescue actions and the protection of personal security were prioritized over border control and national and regional security. Briefly put, ICT-based border surveillance underscores and creates rather than alleviates personal risks as it is employed in migration governance. Casualties as well as the trafficking business at the Mediterranean Sea could, if not be avoided completely so at least significantly reduced by lifting the carrier agreement, by issuing humanitarian visa etcetera. In the following, EU migration governance will be related to aspects that contribute to the need to migrate and to current mass displacement. Responsibility for the situation of displacement are discussed from the perspectives of interdependence and interconnectedness. Having disclosed implications of technology-mediated border control and EU migration governance, the next step will be to investigate responsibility in relation to transnational migration by means of Imamichi's view on interdependence.

4 Interdependence and Interconnectedness in a Globalized World

Imamichi understands the role of ethics as a study of "betweenness" and the (many) relationships of man. He argues that we need to consider the changes in our ontological structure and the multipolarity of relationships today (Imamichi 2009: 11) and he underscores that human beings are dependent, not only on each other, but on nature. Here, the meaning of these two types of dependencies will be illustrated by means of forced, transnational migration. The first interdependence to be considered in relation to mass-displacement is that between man and nature.

As already noted, across the globe, migrants lose their lives in desperate attempts to obtain decent living conditions. The dependence on nature, emphasized by Imamichi, is clearly visible in relation to transnational migration (and internal displacement), past and present. Migration is a classical response to personal insecurity, closely related to poverty and poor living conditions. Typically, the more directly individuals rely on the environment and on natural resources for their living/subsistence such as nomadic cattle breeders, subsistence farmers and small-scale fishers, the more harmed they are likely to be by environmental catastrophes and degradation. If no longer possible to live out of natural resources, individuals may have little or no choice but to seek better living conditions elsewhere. If only certain regions of a nation are negatively affected by environmental change, individuals from these areas may be able to resettle elsewhere within the country (internal displacement), typically giving rise to urbanization, but the larger the affected areas and number of individuals are, they may have to move abroad (transnational migration). Throughout history, people have abandoned settlements for environmental reasons in regions affected by droughts, floods and crop failure. The reasons behind historical environmental migration have foremost been natural but also man-made such as in the case of the US Dust Bowl emigration wave in the 1930s. Already today, millions of people are displaced due to rising sea levels, changes in rainfall, glacial melting, ocean level rise and extreme weather (IPCC 2007). For example, Bangladesh is suffering from glacial melting in the form of flooding and landslides (McFerran 2007). Typically, individuals who, at present, are nationally or internationally displaced for reasons of environmental change are foremost from developing countries.

According to some prognoses, the ratio of individuals fleeing from their home regions due to climate change will accelerate fast. Fears have been voiced that climate change will trigger historically unprecedented waves of mass migration (Fritz 2010) with numbers ranging from 50 to 200 million displaced persons in 2050 (Brown 2007). For instance, the non-profit organization the Environmental Justice Foundation estimates that 150 million persons will have to leave their home countries during the coming 40 years.[10] In 50 years, climate change is said to be the most common reason behind migration (Bell 2004: 139). One way of stressing the urgency of climate change is the denomination of (the controversial term) "climate refugees."

Certainly, the case could be made that the causality regarding environmental change is unclear and that the consequences of climate change are hard to address for the reason that current knowledge of ecosystems and the impacts on them of our actions is incomplete. However, the mere likelihood that human conduct may have significant, adverse effects on the environment speaks to a responsibility to as far as possible seek to foresee, to prevent, to mitigate and to adapt to the negative effects of our "environmental footprints" or imprints. Despite the many uncertainties

[10] http://ejfoundation.org/campaign/climate

surrounding prognoses regarding future climate-induced migration, most certainly, the magnitude and frequency of global warming effects will influence future migration patterns (Fritz 2010). Already under existing conditions, most developing countries face urbanization and in developing nations, large numbers of individuals are internally displaced and fleeing their home countries for climate-related reasons such as drought and flooding. Moreover, climate change may lead to scarce resources – a classical reason for migration and a common component in (armed) conflicts. Even if there are no simple causal links between climate change and armed conflicts, climate change is one of many factors that can affect the risk for armed conflict (Mobjörk and Simonsson 2011) and intensify already existing conflicts if not triggering novel conflicts. Clearly, migration is intimately connected with the environment and the relation man – nature ought careful consideration.

Even if the range and scope of climate-induced migration cannot be established with perfect accuracy, the impact will most likely be strong enough to give rise to largescale emigration. Although climate change is likely to have a worldwide negative impact, already underprivileged regions and the globally worst-off are likely to be most vulnerable to environmental change. That is, environmental change implies a novel twist on existing forms of inequity and maldistribution. Importantly, two major differences between present-time and historical environmental migration can be noticed: (1) the reasons behind environmental change seem to be connected to human activities and (2) the magnitude and frequency of such changes. Manmade climate change has demonstrated an urgent need for shared responsibility and joint action in a matter that clearly stem from complex patterns of interaction, that have global consequences and that are beyond the capacity of single nation states to rectify/address in a just as well as in an effective manner. How such responsibility should be distributed, according to what principles and for what reasons have been debated though.

Human beings do not only depend on natural resources and climate stability, they rely on each other and are interconnected. From this interconnectedness, certain moral obligations are taken to follow. At present, the EU witnesses how large numbers of individuals in despair seek access. War, political instability and persecution trigger mass exodus from regions like Afghanistan, Iraq and Syria and combinations of conflicts and uneven resource distribution force large groups to leave their home countries in sub-Saharan Africa. Just like with environmental change, in an interconnected and globalized world, neither is the cause, nor are the effects, of forced displacement local. In addition to altered environmental conditions, poverty war, and political instability are examples of highly complex forcing conditions. Transnational migration is, in many cases, a reaction to untenable living conditions and poverty (frustrated basic rights) and an attempt to obtain personal security, stability and economic betterment abroad. Untenable living conditions in one country are often, at least partially, explained by involvement of external agents (such as multinational corporations) and other nations. Existing nation states can be seen as results of past and present interactions with and interdependencies between foreign countries and the legacy of colonialism is important for present discussions on migration and borders. Rather than viewing current polities and their characteristics e.g. affluent/poor, stable/unstable as given, existing nations should be seen as results of past and present interactions with and interdependencies between foreign countries (Pogge 2008). Colonialism has contributed to the wealth of some nations and the poverty of others. Not seldom, the wealth of present day developed nations stem from extraction and processing of resources of developing nations. Such conduct has also contributed to the latter's underprivileged situations. Despite struggles for and processes of independence and national liberation, the uneven relationship between former European colonial powers and their former colonies remains. In many parts of the world, national borders have been drawn by European colonial powers. At independence from colonial power, states have been created, often without due respect for ethno cultural, geographical and ecological realities. Some of the consequences are tensions between the many different ethnic groups (tribes) with different historical traditions, cultures and languages now sorting under the same national administration, that the construed national borders disrupt traditional mobility patterns. Present regional conflicts, as well as conflicts over migration and the rights of migrants should also be seen in this historical light. Moreover, the legacy of colonialism is also important for challenging present restriction of migration to Europe. According to a principle that those who contributed to the wealth of a nation also have a right to the results, and the factual claim that colonialism benefitted the economic development of the colonial nation, even descendants of former colonized peoples have a right to the results (Ypi 2009). This is an argument for a generous immigration policy because it implies that the colonized peoples who contributed to the wealth of the colonial nation are entitled to immigrate to get their fair share of this wealth. Having presented ways in which individuals are dependent and interrelated and the relevance for present-time displacement, the next step will be to discuss moral responsibility for those displaced and what that entails for migration governance.

5 Transnational Migration and Shared Responsibility

In the previous section, a rather dystopic scenario regarding future migration was presented, implying an increasing rather than decreasing flow of migrants for climate-related reasons during the next 50 years (unless climate mitigations are significantly more effective than expected). Environmental change will make up another factor triggering migration in addition to traditional factors that continuously will force individuals to relocate such as political instability and different forms of injustice. It comes clear that migration, just like climate and environmental change, is a result of complex patterns of interaction that first and foremost harm the most vulnerable members of the "global village." In order to ensure robust and just solutions, the case will be made that transnational migration must be addressed jointly at a supranational level. It will be argued that the "better-off" have certain duties towards the "worst-off." Inspired by the climate change discourse within which a model for fair "burden sharing" has been developed, a model for equitable sharing of responsibility for migrants is drafted.

In the case of climate and environmental change, criteria for equitable burden sharing have been developed through the many summits on Global Climate Change (GCC), starting with the adoption of the Climate Change Convention during the Rio Earth Summit in 1992.[11] Even if, historically, greenhouse gas (GHC) emissions (primarily) stem from industrialized nations, climate effects can appear anywhere on the planet, irrespective of their source (Intergovernmental Panel on Climate Change, IPCC 2007).[12] Sustained alterations of climate patterns worldwide are so profound and fundamental in nature that nation-states cannot, in an effective manner, tackle CO_2-induced environmental damage on their own. In particular, this is true about the disproportionately harmed and resource-poor developing countries. Developed nations that (historically) have produced the largest share of GHC are now held responsible for the harms caused and for the situation of the worst-off. In order to secure sustainable development, principles for equitable burden-sharing regarding climate change have been developed, prescribing Common but Differentiated Responsibility. "*In view of the different contributions to global environmental degradation, states have common but differentiated responsibilities. The developed countries acknowledge the responsibility that they bear in the international pursuit of sustainable development in view of the pressures their societies place on the global environment and of the technologies and financial resources they command* (Article 3.1)." The CBDR-clause requires that developed nations take on a greater burden in combating climate change than developing countries, which are particularly "vulnerable to the adverse effects of climate change" and "bear disproportionate or abnormal burden under the convention (Article 3.2)." Within the GCC discourse, affluent nations that contributed to GCC and have the capacity to, if not prevent climate change so at least minimize further harm, been considered morally responsible for alleviating negative consequences thereof i.e. they have caused harm and they have the capacity "to do good."

In a similar vein, the case can be made that mass-displacement, just like climate change, is a global concern and a phenomenon of a dignity that is beyond single nation states to address in a potent manner. Both reasons for and consequences of transnational migration are of a magnitude that motivates a shared approach to address the problem. A shared and principled approach is necessary for reasons of justice as well as for reasons of effectiveness. S*tructurally*, international displacement and transnational migration resembles the global climate change issue (and in part, environmental change is a reason behind emigration waves.) Migration is to a large

[11] The United Nations Framework Convention on Climate Change (UNFCCC) is an international environmental treaty negotiated at the United Nations Conference on Environment and Development (UNCED), Rio de Janeiro, June 3–14, 1992. The objective of the treaty is to "stabilize greenhouse gas concentrations in the atmosphere at a level that would prevent dangerous anthropogenic interference with the climate system." *The United Nations Framework Convention on Climate Change*.

[12] http://ipcc.ch/

extent a reaction to untenable living conditions that have been brought about by complex interactions. In the previous discussion on interdependence and interconnectedness it was shown that developed nations in different ways have contributed to the present situation of mass-displacement and the influx of migrants to stable regions such as the European Union. They have contributed and continuously contribute to untenable living conditions in developing nations that trigger emigration. From this involvement and impact, a certain moral responsibility can be said to follow. "*If we have contributed to the predicament of foreigners, we are morally responsible for their situation in the same way we would be if fellow nationals would suffer from our actions* (Pogge 2008)." Taking responsibility for causes and effects of transnational migration is a moral obligation that receiving nations *owe* migrants due to a certain causal involvement in the affairs of sending countries, triggering transnational migration. Moreover, it can be argued that the EU is responsible both for deprivations that enforce emigration and for placing restrictions on underprivileged individuals' chances to balance the harmful effects imposed on them by seeking admission and personal security in stable regions such as the EU. That is, the reason why third country nationals migrate in the first place, can often be traced to causal involvement of affluent nations and transnational organisation. For these reasons, the privileged members of the global village should engage in equitable responsibility sharing for transnational migration. Nations or regions that contribute to migration should take responsibility for the causes of transnational migration and the migration flow. Not only do developed nations bring about emigration, they seek to prevent irregular migrants from entering their territories, and from exercising their fundamental human right to seek asylum abroad, for instance by means of strict admission criteria and by encompassing border monitoring programs. In consequence, migrants are forced to relocate in an irregular fashion and as irregular migrants they exposed to significant risks. As already argued, the frequency and magnitude of the migration-related tragedies at sea stem, to a significant extent, from a securitization of borders (Stumpf 2006) and from increasingly strict immigration policies within the privileged, developed world (Pécoud and de Guchteneire 2006). Although the causal chains are complex, and responsibility for specific events hard to establish, the involvement of developed nations and transnational organisations is significant enough to speak of a moral responsibility. Moreover, resourceful developed nations should take a larger responsibility for forced migration than the developing nations most affected by emigration as well as immigration. In particular, affluent nations should seek to prevent conditions that trigger migration rather than adopting reactive measures to existing migration flow such as strict border- and immigration control. In the same way as with Global Climate Change, states should be seen as having Common but Differentiated Responsibility (CBDR)[13] with regard to displacement and migration.

---

[13] The CBDR principle is one of the main pillars of sustainable development, drawing on equity considerations and explicated in the 1992 Rio Earth Summit.

6 Conclusion

Despite the fact that the largest migration flow goes between developing nations and even if the greatest numbers of displaced individuals reside within developing countries, the philosophical migration-debate has typically been concerned with the effects of irregular migration on developed nations and their conditions and interests. The debate has revolved on welfare-states' rights and duties *vis-á-vis* universal rights becoming migrants e.g. freedom of movement. In practice, territorial security of affluent nations has typically been allowed to trump individual rights to personal security and protection. Practices of exclusion and border protection have typically been justified by reference to the sovereignty principles and that nation states primarily have obligations towards members, not towards nonmembers. Little attention has been paid to root causes behind transnational migration and to the effects of migration on developing countries. In order to better understand and address the highly complex problem of international displacement and mass migration, inspiration has been sought in Imamichi's eco-ethica. His critical stance to the usage of technology and obligations stemming from instrumented-action serve as a starting point to discuss the ethics of ICT-mediated border and immigration control within the EU. The claim made here is not so much that ICT gives raise to ethical problems but rather that the effectiveness of ICT-based border control sheds light on the inhumaneness of the current EU migration regime. Following Imamichi, there seems to be an imperative to make as good use as possible of technological advancement. If human rights can be promoted and if harm and casualties can be avoided by means of technical instruments/instrumented action, there is an obligation to do so. However, employed in border governance, ICT is used to enforce unjust immigration directives, rather than fulfilling its potential to save lives. If all states would be equally resourceful as the EU is and could protect their borders as effectively, refugees would have nowhere to turn for protection. By amplifying the effects of current technology mediated border control, it comes clear how problematic immigration policies are. Furthermore, Imamichi's holistic approach and his emphasis on interdependence are used to reflect on factors that trigger displacement, responsibility for and proper responses to transnational migration. Imamichi's view on interdependence between man-nature and between man-man, and broader notion of responsibility has been used to gain a new perspective on the ethics of technology-mediated action in the context of EU migration governance. First, it has been argued that transnational migration, just like environmental change, is a global concern and that developed countries' are (at least) co-responsible for untenable living conditions in underprivileged regions of the world, and consequently for migration that is intimately connected with poverty, instability and unequal living standards. It has been demonstrated that international displacement and transnational migration are closely linked to the environment and that altered environmental conditions often are mirrored in migration patterns. Given the prognosis for climate change, transnational migration is likely to increase rather than decrease. It has also been illustrated that developing countries have, even if the causal links are rather blurred, been involved and are involved in activities that contribute to untenable living conditions in terms of poverty and inability that trigger emigration. Second, the case has been made that ICT-based migration governance is complicit in the creation and sustainment of an unjust, unsafe and unsustainable world. Migration governance is one domain within which sophisticated surveillance (capable) technology could be employed to, if not build away and completely eliminate casualties, so at least reduce risks and save many lives. However, thousands of "untimely deaths occur annually, and in many cases, ICT-based migration control increase rather than reduce risks associated with forced migration. To address these problems, supra-national solutions are needed, for reasons of justice as well as for reasons of effectiveness, and an equitable responsibility sharing for transnational migration has been advocated. Both reasons for and consequences of transnational migration are of a magnitude that motivates a shared approach, including sending and receiving countries. Unauthorized migration is to a large extent a rather desperate reaction to untenable living conditions that have been brought about by complex interactions. And, in order to be just and effective, solutions must focus root causes rather than, as today, mere symptoms.

References

Ajana, Bthaj 2015. "Augmented borders: Big Data and the ethics of immigration control", *Journal of Information, Communication and Ethics in Society* 13.1:58–78.

Brian, Tara and Frank Laczko. 2014. "Fatal Journeys, Tracking Lives Lost during Migration, International Organisation of Migration Report", *International Organization of Migration Report* http://publications.iom.int/bookstore/free/FatalJourneys_CountingtheUncounted.pdf.

Britz, Johannes. 2013. "To Understand or not to Understand: A Critical Reflection of Information and Knowledge Poverty" *Information Ethics in Africa: Cross-cutting Themes*. Pretoria: ACEIE: 71–80.

Broeders, Daniel. 2007."The New Digital Borders of the European Union" *International Sociology* 22: 71–92 doi: https://doi.org/10.1177/0268580907070126.

Broeders, Daniel and Godfried Engbersen. 2007. "The Fight Against Illegal Migration" *American Behavioural Scientist* 50: 1592 DOI: https://doi.org/10.1177/0002764207302470.

Brown, O (2007) Climate change and forced migration: Observations, projections and implications, Human Development Report 2007/2008 Fighting climate change: Human solidarity in a divided world, Human Development Report Office, OCCASIONAL PAPER Available on November 12th at: https://www.iisd.org/sites/default/files/publications/climate_forced_migration.pdf

Carter, Robert E (2001) *Encounter with Enlightenment – A Study of Japanese Ethics*, New York: State University of New York Press.

Dijestelbloem, Huub. 2009. "Europe's new technological gatekeepers. Debating the employment of technology in Migration Policy, *Amsterdam Law Forum* 1.4:11–18.

Dijstelbloem, Huub and Albert Meijer. 2011. *Migration and the New Technological Borders of Europe* (Migration, Minorities and Citizenship), Basingstoke: Palgrave Macmillan.

Fritz, Carolina. 2010. "Climate Change and Migration: Sorting through Complex Issues without the Hype" *Migration Policy Institute Feature* Washington, DC: Environmental Justice Foundation, 6:18–19.

Gündoğdu, Ayten. 2015. *Rightlessness in an Age of Rights: Hannah Arendt and the contemporary struggles of migrants*, Oxford: Oxford University Press.

Gorniak-Kocikowska, Kryztyna. 1996. "The Computer Revolution and the Problem of Global Ethics" in Terrell Ward Bynum and Simon Rogerson, eds., *Global Information Ethics*, Opragen Publications:177–190.

Hernández-Carretero, María. 2009. "Reconciling Border Control with the Human Aspects of Unauthorized Migration" *International Peace Research Institute, Oslo (PRIO)* http://www.prio.no/sptrans/724731520/Reconciling-Border-Control.pdf.

Huysmans, Jef. 2006. *The Politics of Insecurity: Fear, migration and asylum in the EU*. London and New York: Routledge.

Imamichi, Tomonobu. 1984. "The Concept of an Eco-Ethics and the Development of Moral Thoughts" *Man and Nature* ed. George McLeon, New York: University Press of America (1984).

Imamichi, Tomonobu. 2009. *An Introduction to Eco-ethica*, Lanham, Maryland: University Press of America.

IPCC, 2007: Climate Change 2007: Synthesis Report. Contribution of Working Groups I, II and III to the Fourth Assessment Report of the Intergovernmental Panel on Climate Change [Core Writing Team, Pachauri, R.K and Reisinger, A.(eds.)]. IPCC, Geneva, Switzerland, 104 pp.

Johnson, Deborah. 1999. *Computer Ethics in the 21st Century* Keynote address at ETHICOMP99, Rome, Italy, October. Published in Spinello, Richard A and Herman H Tavani (eds.) Readings in Cyberethics. Jones and Bartlett, 2001.

McFerran, Ann (2007). 'Bangladesh: A Nation in Fear of Drowning," The Independent, 2007, Available at http://www.independent.co.uk/environment/climate

Mobjörk, Malin and Louise Simonsson. 2011. "Klimatförändringar, migration och konflikter: samband och förutsägelser," FOI-R%2D%2D3297%2D%2DSE.

Palm, Elin. 2014. "Towards Equitable Responsibility-sharing for Forced Migration" *International Journal of Migration and Border Studies* 1.3:253–275.

Pécoud, Antonine and Paul de Guchteneire. 2006. "International Migration, Border Controls and Human Rights: Assessing the Relevance of a Right to Mobility", *Journal of Borderlands Studies* 21.1:69–86.

Pinyol-Jiménez, Gemma. 2012. "The Migration-Security Nexus in Short: Instruments and actions in the European Union". Amsterdam Law Forum, VU University Amsterdam http://ojs.ubvu.vu.nl/alf/article/viewFile/255/442.

Pogge, Thomas. 2008. *World Poverty and Human Rights: Cosmopolitan Responsibilities and Reforms*, 2nd ed. Cambridge: Polity Press.

Russell, Bertrand. 1948. *History of Western Philosophy*, London: George Allen and Unwin Ltd.

Stumpf, Judith. 2006. "Crimmigration Crisis: Immigrants, Crime, and Sovereign Power", *American University Law Review*, 56.2:367–419.

Ypi, Lea. 2009. "Associative Duties, Global Justice, and the Colonies" *Philosophy and Public Affairs* 37.2:104–135.

Chapter 11
Social Media in a Disaster: Technology, Ethics and Society in Tōhoku in March 2011

Asai Ryoko

1 Introduction

On March 11th 2011, Japan was hit by an undersea megathrust earthquake of magnitude 9.0, which is referred to as the 2011 Tōhoku Earthquake, the 3.11 Earthquake or, in Japan, the Great East Japan Earthquake (*Higashi Nihon Daishinsai* 東日本大震災). An extensive geographic area in Japan's Tōhoku 東北 (northeast) region was seriously damaged; especially the coastal region was totally devastated by tsunami. Everything was engulfed by enormous tsunami. Residents, houses, high buildings, schools, vehicles and almost all objects vanished away from the land. As of June 2017, the Japanese national police agency announced that the number of dead was 15,894, and that 2550 people were still missing (National Police Agency 2017). The disaster disrupted almost all public utilities, including electricity, gas supplies, water, radio, telephone, TV and Internet systems. In the face of the massive damage and aftershock, people utilised different kinds of media in order to know what had happened to them, their communities and the country.

In the damaged area, however, almost all media were ruined by the disaster. It was very hard for the disaster victims to contact others and to obtain information about their situations via telephone, TV or other existing media. Though victims barely escaped with their lives due to the enormous tsunami, many of them had mobile phones (including smartphones) in their pockets, and Internet access were still avail-

The original version of this chapter was revised: MS Mincho font was updated throughout the book for Japanese and Chinese characters. The correction to this book is available at https://doi.org/10.1007/978-3-319-59027-1_13

Asai Ryoko (✉)
Department of Information Technology, Uppsala University, Uppsala, Sweden

Centre for Business Information Ethics, Meiji University, Chiyoda, Tokyo, Japan
e-mail: ryoko.asai@it.uu.se

Thomas Taro Lennerfors, Murata Kiyoshi (eds.), *Tetsugaku Companion to Japanese Ethics and Technology*, Tetsugaku Companions to Japanese Philosophy 1,
https://doi.org/10.1007/978-3-319-59027-1_11

able in some parts of the damaged area. Immediately after the big earthquake happened, most of disaster victims were stranded and completely cut off from non-disaster areas. However, after a while, those victims who evacuated to a safer distance from the most severely damaged area attempted to contact with their missing loved ones and retrieve information about them with limited devices, battery power and electricity supply. In the meantime, people living in light damaged and non-damaged areas tried to send various kinds of information via the Internet to support victims.

Soon after the terrible disaster, people, including both victims and non-victims, started to use social media actively as a means of communication and information tools, and exchanged a large amount of information through social media such as Twitter to restore or maintain social ties. On the other hand, however, ethically questionable social media uses such as wilful or inadvertent spread of disinformation and misinformation were often come across. This study explores how people use social media under serious social conditions and how social media influence on people's behaviour in the face of a disaster from the viewpoint of information ethics, based on the case of the 2011 Tōhoku Earthquake disaster in March 2011.

2 Information and Communication Technology as a Daily Necessity

Japan is one of the most advanced information and communication technology (ICT) countries in the world. Japanese people already enjoyed fast and relatively low cost Internet access in March 2011 when the earthquake hit. According to the Japanese government white paper of 2010, the total population of Internet users was estimated at 94.08 million, the number of social networking service users stood at 71.34 million and the number of bloggers to 26.95 million (Ministry of Internal Affairs and Communications 2011).[1] As of 2006, the economic scale of social networking services was estimated at 56.8 billion JPY and blogs at 16 billion JPY, and the market scale of Japanese social media was estimated in total to be about 196.1 billion JPY (Ministry of Land, Infrastructure, Transport and Tourism 2006). It is not hard to imagine that the economic scale of social media and its importance in Japanese economy has been increasing more and more these years.

Why did so many Japanese join social media which together created a significant economic market? One of the main reasons is the low cost of the Internet. Broadband Internet infrastructure was established almost all over Japan by the late 2000s and a large number of users enjoy optical fibre access at a low cost. This means that people have high-speed access to the Internet. Japanese people can use the Internet as one of the most important daily necessities. Before the 3.11 earthquake, many peo-

[1]According to the Japanese government white paper of 2016, the number of Internet users in Japan is around 100 million people and it accounts for 83.0% of the population in fiscal 2016 (Ministry of Internal Affairs and Communications 2016). However, this study refers to the data of 2010 in order to explore ethics and social media use at that time when the 2011 Tōhoku Earthquake happened.

ple had already joined social media and communicated with others through it, and they accessed the Internet through mobile phones as well as personal computers (Ministry of Internal Affairs and Communications 2011). Therefore, they could get and exchange information whenever they wanted, even when they were walking in a street, getting on a train or talking with friends.

Novel and advanced technology greatly impacts on our daily-life environment. It is needless to say that new technology dramatically enhances the convenience in every aspect of our daily lives. And also, it greatly affects the quality of experiences in our lives. According to Japanese philosopher Ichikawa Hiroshi 市川浩, who is well-known for his theory of the body, the development of technology alters our experiences, which means how we get experience, perceive the reality where we live, and establish the relationship with others in our lives are changing along with technology development (Ichikawa 1992). Basically we get experiences subjectively, such as seeing something firsthand, talking with others in person, touching something directly etc. However, high-tech environment could provide us with more opportunities to experience something regardless of time difference, geographical distance, national boundary and so on. Ichikawa shows two different kinds of experiences in our lives; first, we can gain an experience in person (firsthand experience or *chokusetsu-keiken* 直接経験); second, we can get an experience mediated by tools, machines, digital signals or symbols (indirect experience or *kansetsu-keiken* 間接経験), especially an experience via symbols is defined as informational experience (*jōhōteki-keiken* 情報的経験).

The development of ICT not only gives people more and more opportunities to have informational experiences, but also increases the value of informational experiences. Although the act of looking at or reading information about things with our own eyes in front of a monitor looks like firsthand experience, there is no subjective or self-controlled experience in the situation.[2] Ichikawa called this kind of experience pseudo-firsthand experience (*gijiteki-chokusetsu-keiken* 擬似的直接経験). Experiences in our lives are elaborated by the mixture of firsthand experiences, indirect experiences and pseudo-firsthand experiences (Ichikawa 1992). People place more value on information and informational experiences, and change the way of thinking and living these days. This change makes it more difficult to recognise clear boundaries between these three kinds of experiences. And the technology development allows people to get an experience easier and faster than before, whereas the living environment becomes more open and complicated for them. Moreover, the technology development makes the living environment artificial and superficial, and the way to perceive others and the environment is changing. In such an artificial and superficial living environment, it is hard to realise the deeper or hidden meaning of things (Ichikawa 1992).[3] In other words, experiences increasingly lose their depth, and, nevertheless, the desire to have deep relationships between things and our existences is increasingly enhanced.

[2] For example, we read traffic accident news and see pictures showing the accident site online with our own eyes, but the news doesn't give us any direct experience of the accident.

[3] In this context, the word "things" includes not only material things which you can physically access/touch, but also non-material things such as actions, phenomena etc.

There is another possible reason why social media attract users strongly. Social media can enrich informational experiences, and provide us with a great opportunity to share our firsthand experience with others and to have an indirect experience and a pseud-firsthand experience through other's experience. Furthermore, the active use of social media could fulfil the desire to have someone who listens to us, shares experiences with us, and confirms our existence.

3 Social Media Use at the Time of Disaster in Tōhoku in March 2011

3.1 "Hand-Made" Information on Social Media: The Case of "Person Finder" by Google

The 3.11 earthquake and following giant tsunami suddenly hit a large geographic area of Japan's Tōhoku region. Directly after the earthquake, existing media such as newspapers, radio and TV did not work well in the damaged area due to the disruption of lifelines. Many journalists tried to report on what had happened to the damaged area. However, almost all roads to that area were cut off and it was very difficult to enter the area physically. The government prohibited the entry to particular parts of the area because of the high level of radiation that had leaked from the Fukushima Daiichi Nuclear Power Plant after the tsunami attack. Soon after the disaster, even the central and local governments had little information about the level of damage and the number of victims in the area of distress. In the social chaos, people had no tools either to find out about the safety of their families, relatives and friends, or to get information about what had happened to Japan.

However, even under this very difficult situation, in the damaged area, the Internet still worked and people used it in order to check on levels of safety and gather information about the disaster by using emails and social networking services. Many private organisations offered a wide variety of assistance to disaster victims. For example, the global ICT company Google set up more than thirty kinds of online services within 24 h after the disaster. Among different services, satellite pictures and maps were provided to rescue teams, the government, international non-governmental organisations and individuals, and contributed to rescue victims and support them. Google's "Person Finder" (Fig. 11.1) was greatly used by people who wanted to find missing people and check on whether their families, relatives and friends were safe (Editorial Office of Computer Technology 2011).[4]

Google placed "Person Finder" on Google Apps immediately after the Japanese earthquake, using the scheme which had improved through some experiences of other disasters in the past. "Person Finder" became the most useful social media platform in Japan at that time in order to confirm people's safety. It was constantly

[4] http://google.org/personfinder/global/home.html (Accessed 18 July 2017)

Fig. 11.1 Google's "Person Finder" at the time of the 2011 Tōhoku Earthquake. (Source: https://www.flickr.com/photos/ramblinglibrarian/5546642234/. Accessed 21 July 2017)

updated through cooperation with the Red Cross and the Nippon Hōsō Kyōkai (NHK; Japan Broadcasting Corporation).[5]

However, in severely damaged areas, most of the conventional media were destroyed by huge tsunami waves. The many people who were stranded did not have any kind of tools to tell and/or find out about the safety of their families and friends. How did they register their safety on the Internet without any digital tool? They wrote messages on paper cards and put the cards on walls in a refuge where they evacuated. However, it was very difficult for them to send messages to people outside the refuge.

Every refuge had message boards on the wall, and the evacuees wrote and posted message cards that included their names or other private information. In some areas, a few people accessed the Internet via mobile phones for a short time. They took pictures of message boards by using mobile phones and uploaded these to Picasa pictures websites.[6]

Many volunteers, who were living in areas that were not much seriously damaged, checked on those pictures through computer monitors, wrote down every

[5] This corporation was the only Japanese public broadcasting company involved in finding and checking on people in the damaged area.

[6] Google provides the photo service Picasa since 2002: http://picasa.google.com/intl/en/ (Accessed 21 July 2017). And also they offer "Google Person's Finder" service since 2010: https://google.org/personfinder/global/home.html (Accessed 21 July 2017). In terms of message boards with handwritten message cards at a refuge in the Tohoku earthquake, see, for example, http://www.google.org/crisisresponse/kiroku311/chapter_02.html or https://blogs.yahoo.co.jp/kfuji_taxi/62119114.html (Accessed 21 July 2017).

message on paper by handwriting and posted those messages on "Person Finder" instead of the victims themselves.

Nowadays, many companies promote a "paperless" policy positively, and ICT is a critical tool for improving their corporate performance. However, when we face the lack of electricity and data access tools under a social crisis, we cannot avoid to reconsider the social role of ICT and the way how to choose proper information, and also to take more notice of what is information.

Google's "Person Finder" was constructed and used not only by new digital technologies but also by conventional analogue manpower. It helped greatly people find families and friends in a chaotic situation after disaster. On the other hand, "Person Finder" evokes a privacy issue. People uploaded huge information about victims, including their names, address, safety status, and pictures to find them as soon as possible. In a critical situation, we need some clues in order to find missing or separated people and check on their safety. Every tiny clue is required to judge the authenticity of the situation. However, there is a social risk of violating people's privacy, sometimes without noticing. Once information is posted on a website, it could remain on the Internet for a very long time-period, due to its survivability. In finding victims' safety as well as specifying victims, the exposure of private information might work well.

3.2 Collaboration Between "Old" Media and "New" Media

The massive disaster in March 2011 damaged the coverage of mass media because it destroyed TV towers and broke telephone cables in the devastated area. For example, TV news programs continued to broadcast live reports not only by using conventional broadcasting systems but also through cutting edge technologies. In other words, existing media collaborated with "new" media in sending information to an audience. In particular, TV companies distributed their TV programs on the Internet at the same time that they broadcasted on TV. TV companies and big Internet companies collaborated with each other to enable a fusion between TV and social media. The TV reported the original news accompanied by people's tweets from Twitter on the TV screen, and social media delivered not only information from users but also streamed live videos from TV programs to their websites.

Thus, people living in the non-disaster area watched real-time events through mass media and social media, and got to know what happened at that time. However, although people living in a damaged area could somehow send a little information by mobile under very limited access and low battery conditions, they could not get enough information to know the situation where they faced after the disaster through mass media. And many of them were stranded in the middle of disaster area without any information from outside. But still people tried hard to help victims by using what they could use at that time.

In this situation, social media complemented the absence of mass media with its communicability and transmissibility, with cooperation and collaboration between mass media and the Internet. Many people used social media like Facebook or Twitter, and some ICT companies played great roles in distributing information using their technologies. As already mentioned, Google created a special website called "Person Finder" to find and check on people's safety in collaborating with NHK. Twitter set up a hash-tag in order to gather information efficiently (Editorial Office of Computer Technology 2011). Under limited Internet and electricity availability, twitter's short and simple but purposive message worked better in sending information to victims. And, some Internet server business companies provided many mirror sites for users. However, the fusion between mass media and Internet did not last not for long. A few weeks later, it ended and both media again delivered information separately just as they had done before the disaster. On the other hand, it became a great opportunity to consider how media should/could work and also what is required for media in a disaster.

People, including both victims and non-victims, exchanged much of information through social media, in order to survive and to support others. Twitter was especially useful to inform where victims were and who needed to be rescued. Twitter contributed greatly to exchange information after the disaster. It was because of its uniqueness: quick reporting, convenience (simpleness), propagation (transmission capability) and diversity (Editorial Office of Computer Technology 2011). Twitter's character limit (140 characters) was effective to make tweets clearer and easier to be read by others in a time of disaster. And also Japanese writing system is able to have a higher information content per character than western languages. When users write the maximum number of characters in a tweet, a tweet written in Japanese is contained 1.4 times the amount of information compared to an English tweet (Neubig and Duh 2013). Because of these reasons, twitter was useful to send more information by simple and short message during the disaster. Thanks to the collaboration between TV and social media, people got a broad range of information about the disaster through the TV news while they checked on the reality and local information through Twitter in a real-time fashion. In the disaster where there was a very limited availability of information tools in their hands, social media became widely known as one of the most useful communication tools and the number of users grew.

Some time after the disaster, many users selected information which was more useful and suitable for their own situations among the large amount of information, and also information focusing on specified or local areas was increasingly sent. And they relied on tweets about local information. This is because users had a certain level of media literacy, and they could not help but choose information that seemed to be plausible based on their literacy as proper one among the flood of information. More importantly, people shared a sense of trust in others in Japan enough to believe unproven information brought by an unknown person during the disaster (Editorial Office of Computer Technology 2011; Kobayashi 2011).

3.3 Consequence of Altruistic Messaging

Under a crisis situation, many Japanese people believed that it was natural to help and support each other, based on the Japanese traditional virtue of "*nasake wa hito no tame narazu* 情けは人の為ならず" which means that sympathetic and supportive attitude is done for oneself, not for others. In other words, the good you do for others is good you do for yourself, or what comes around goes around (*inga ōhō* 因果応報). Soon after the disaster, social media were used by many people in order to support victims and help each other. Mass media only provided general information. On the other hand, people suffered from the disaster needed local and detailed information, such as specified area's traffic information and the place available to buy water bottles. It was viral information that met those local-oriented needs among victims. We saw a lot of occasions where people said "let us stand by with each other in time of need (*komatta toki wa otagaisama* 困った時はお互いさま)". Those attitudes and words are reflected the Japanese virtue of mutual support. At first glance, it seems to be self-sacrifice or altruistic. However, the altruistic behaviour clearly shows ethical problems with use of social media in chaotic situations like a disaster.

Basically, in the disaster time, people spread information for the purpose of helping others. However, in a chaotic occasion, there are not only people who try to do something good for others but also people who try to fool others. When many people used social media actively with various kinds of intentions and interests in a certain context, two different types of false information can be flowed on the Internet as well as correct and useful information. One type of false information is categorised as *disinformation* including demagogic information transmitted based on malicious intent, for example, to make a society disordered or confusing. The other is *expired information*, which is spread based on humanity or goodwill, but which, at the time others gets the information, is already useless because of the time-lag involved in sending information (Editorial Office of Computer Technology 2011; Hayashi and Yamaji 2012; Kobayashi 2011). And both types of false information could lead to misinformation when people believe in false information. Among two false information types, the latter one, *expired information*, contains profound ethical issues.

In the case of expired information, information senders act based on goodwill (at least the motivation to do something good for others). And the altruistic act is rational for themselves (Nagel 1970). However, information sent by those senders do not always contribute others, rather it might cause mess. For example, unnecessary ambulance call, or unbalanced aid supplies were often seen in the 2011 Tōhoku earthquake. It is very difficult to accuse those senders of being irresponsible one-sidedly. Results, whether they are favourable or unfavourable, were consequences of people's goodwill or altruistic act. Who does take responsibility for consequences in the end? When it comes to goodwill senders, they get satisfied after sending message to others and they believe they could do something good for others. In this case, the consequence is the fact that they get satisfied. And mostly they do not follow up what information causes or how their information works for others. When we focus on the consequence of the spread of such information, or what such information cause to others in the end, those message senders do not concern carefully about result of what they do for others and they are not enough aware of being ethical in utilitarian perspectives. On the other hand, when people spread information to help others, like in a disaster time, they believe it should be done and it would be ethical act as a person, mostly without checking if it is true by themselves.

New technology, such as twitter, enabled and encouraged a large number of people to conduct an "ethical act" via the Internet. People living in little or non-damaged area saw the disaster seriously affecting the Tōhoku area of Japan on TV or online. They were then bombarded by information about the disaster via both old and new media. Media gave people indirect experience or pseudo-firsthand experience of the disaster. They knew it indirectly, but realised how the disaster did awful damage to the Tōhoku area and there were many victims waiting for their help in the damaged area. Their indirect experience and pseudo-firsthand experience evoked frustration among them, because they could not do anything for victims from a distance. Social media gave them the way to actively involve themselves in and contribute to the disaster situation, and they tweeted a lot to spread information they believed useful to support victims as an ethical act. Their sending and spreading messages, regardless of whether those messages were correct or not, changed their indirect experience to direct experience, and satisfied them with the sense of doing "good" via social media.

However, no one took any responsibility when the spread information caused bad consequences. Still, such messages, even false ones, were dispatched based on senders' altruistic or moral motivations. Especially in the chaotic situation, there was only a little time for them to think what consequences were led to by their messages and judge whether the information dispatched was correct. Senders had to decide autonomously and quickly what they should do with information they acquired using social media. Therefore, people who intended to help victims were required to have a higher information literacy to correctly judge the credibility and validity of information dispatched.

4 Social Capital and Social Media

After the disaster, one concept concerning tsunami got a lot of attention and became widely known in Japan. The concept is "*tsunami tendenko* (津波てんでんこ)", which asks everyone for "a quick tsunami evacuation without waiting for others, not even one's parents or children" (Yamori 2012). This concept does not mean that only you can survive without caring about others. According to Japanese psychologist Yamori Katsuya 矢守克也 who is working for the Research Center for Disaster Reduction Systems at Kyoto University, *tsunami tendenko* is a mindset in a tsunami disaster and has been passed down from father to son for long years in Tōhoku area (Yamori 2012). It contributes to these three aspects: (1) prompting evacuation by others following the initial evacuees' behaviour; (2) reducing preventable death and injury caused by attempting to help someone who failed to escape; and (3) easing survivor's guilt in the aftermath, for evacuating while others were left behind (Yamori 2012: 35). When we aim to have more survivors in a disaster, this concept is expected to contribute to it greatly. However, in order to exercise the function of the mindset fully, we need to meet a very important condition.

When we evacuate separately from others, even without waiting for others, we need to believe that others like your family, friends and neighbours also evacuate at the same time. This means we need to establish mutual trust each other over time, in terms of emergency evacuation. Such mutual trust is considered as social capital. Conventionally, the function of mutual trust and support used to be carried by a community, including families and neighbours. In the real world, people form real-existing communities around their families or residential areas, where they can exchange their communities' information, or sometimes rumour or gossip, in the streets, supermarkets or somewhere else. People maintained close communications and lived by helping each other through attending community events and maintaining relations among neighbours.

However, participation in community organisations in Japan has decreased since recent years (Ministry of Internal Affairs and Communications 2011). Instead, after ICT grew to be used by many people, they can exchange information, without meeting anywhere in the real world, and can contact others living in other communities through the Internet. Now people can use new media via social media like blogs, social networking services and personal websites. These new media offer a great opportunity to know others and feel for each other socially regardless of where they live. In this social trend, how do we establish social capital?

The development and deployment of ICT have expanded opportunities for us to know others and to communicate with others. In other words, ICT has made it possible for people to establish social relations to many other people. In particular, at present, social media plays an important role in making a place or an *agora* where people communicate with each other. As social media penetrates information literacy, it makes remarkable progress in developing the information gathering-sending abilities of individual users. Some people use social media as a chance to relate to other people in a virtual space.

These phenomena seem to be connected with political apathy and urbanisation under globalisation. Younger generations prefer to avoid a community life because they lead a highly privatised and individualised life. There is a little chance to have an informal social and/or public life in a community (Oldenburg 1999). The absence of an informal public life dilutes human relationships and community ties. In conventional settings, people had learnt the way of living through communication with local community members. That process made it an opportunity to make an informal public life (Oldenburg 1999). An informal public life might be a basis for a social network and develop social capital to enrich life.

How do social media make and complement a social network in order to become a basis for social capital? If people use the Internet only for browsing websites, they cannot form social ties. In order to establish a social ties or relationships, they need to be interactive and communicate mutually between users. Social media offers a huge possibility to form and maintain a social network because of its interactivity. Needless to say, many users gather in online communities. They may never meet each other in their entire lives, or they may not know each other's real faces and real names. Even if people do not know each other at all in real life, they can communicate in an online community beyond geographical borders and time differences through the Internet.

This is a very different characteristic from conventional face-to-face communication. Because of this characteristic, there are possibilities to create new forms of social networks, to change the meaning of trust, or to form different social norms within a community than before. In particular, an online community might function to promote building flat social networks or spontaneous participation. Even when people share a common interest and purpose, they can access and communicate with each other in an online community. In this type of social network, we can see "thin trust" and generalised reciprocity (Putnam 1993a, b, 1995a, b; Lin 2001). These could generate particular norms in a social network, and contribute to reducing the risks of coordination failure and increasing social exchange as social capital (Coleman 1987, 1990).

When people look for information and expect others to offer them useful information, they can post questions directly on social media. They can gather a lot of information and solutions through social media, regardless of its correctness and validity. In the event that a particular user's information contributes to a certain questioner's benefit, this type of information exchange nurtures what is called "thick trust" and specific reciprocity (Miyata 2005). When people send information to many and unspecified others, this information exchange type implies the possibility that social media works as a relatively public-oriented information hub among many users. On the other hand, when they exchange information between specified users, the information exchanged has private characteristics and it reciprocates between particular users. Thus, social media creates two kinds of social networks; one is positioned as a resource of public property; another is considered as a resource of private property. In both types of networks, social media contribute to communication among users and offer the possibility to establish mutual trust. Social media might also lead to foster social capital based on mutual trust. As the case of the use of tweeter at the disaster situation showed, users tweeted to support or help others without any visible reward or reciprocal exchange from others. In this respect, tweets for unspecified many others in the 2011 Tōhoku earthquake could be more public-oriented. However, it is not as if there was no reciprocal exchange between message senders and receivers at that time. Receivers could obtain useful information relating to the disaster, and senders could complement a lack of firsthand experience and be satisfied with their own ethical acts by sending message via social media (see Sect. 3.3). Thus, what senders gained there could be considered as invisible and non-material rewards to promote their altruistic behaviour on social media.

The concept *tsunami tendenko* has been locally handed down from parents to children. But, after the disaster, it has been widely known how the concept works in a disaster time. Needless to say, we can find a lot of information about *tsunami tendenko* not only online news sites but also on social media. People started to know the term and notice how it works little by little. Of course, social media might be just a trigger to let people know it. But social media bring us a chance to talk about it at home, in the community, in the working place and on the Internet, and the conversation with others might open up a new possibility to build mutual trust with others.

Moreover, social media is changing our communication behaviour gradually. When we communicate with others and understand them both in real and virtual spaces, the social context behind information exchanged is very important in interpreting it and understanding the information senders. The social context of communication is defined by geographic, organisational and situational variables, and those variables influence the content of communication among people (Sproull and Kiesler 1991). In order to discern social context cues, communicators observe static cues (such as the physical setting and location) and dynamic cues (including non-verbal behaviour like gesture or facial expression) in communicating with others. Communicators' behaviour is determined based on social context cues, and they can adjust their behaviour depending on situations through the process of interaction.

In online communication, however, it is more difficult for communicators to perceive others' static and dynamic elements as compared to face-to-face communication. In many cases, social media limit the number of characters and the amount of data that can be posted while making it possible for users to communicate regardless of physical distance, national boundaries and time differences (Sproull and Kiesler 1991). Generally, participation is considered as the key element in the recent trend toward democratisation. In fact, numerous users send and receive a huge amount of information via social media so as to cultivate relationships with others and strengthen mutual trust beyond borders. It is recognised that social media advance participation through the exchange of information with minimal social context cues (Asai 2011; Sproull and Kiesler 1986, 1991).

On the other hand, social media is considered as a political artefact (Winner 1995). It could affect social power structures directly or indirectly by its design. Present society locates in the place where dynamic political chaos and technological stability and functionality mix. Especially under the chaotic situation like a big disaster, considering influence of social media on politics and society is needed. That means to deliberate what and whose intention is reflected on and who controls social media including information spread by social media. For example, in chaotic situations, social media could manipulate easily public opinions and information by programming and filtering information when people have little information about their own situation. They have no other way but to trust information which they get. Designing technology mostly focus on universality, sustainability and reparability. However, given ethical use of technology, ensuring fairness, accountability and accuracy is also required in designing and developing technology.

5 Conclusion

In the case of the 2011 Tōhoku disaster, the Japanese shared an extremely strong experience of the disaster. The extreme experience became the shared social context in communicating with others. In the wake of the lack of information available during the disaster period, many people joined social media and started up a communication with others positively. Social media changed their indirect disaster experience to firsthand experience through sending message and spreading information, and let them do something "good" to others. For example, they encouraged each other by exchanging messages posted on social media. In those processes, many messages were very conscious: they emphasised that the Japanese could surely survive, and that Japan would recover from the disaster in the future because of people's solidarity and great efforts. For example, many people tweeted "We are the Japanese, we can overcome the disaster", "We believe in Japan's power", or "*Ganbaro! Nippon!!* 頑張ろう！日本!! (Do the best! Do overcome the disaster, Japan!!)".[7] These messages seem to reflect Japanese cultural characteristics, for instance, the Japanese working virtue as industrious and conscientious workers. People strongly awoke and enhanced their national identity as Japanese. They developed a sense of solidarity through messages encouraging other people (Asai 2011).

It is often said the reason why there are less disaster victims and damages in Japan comparing to other countries is that the Japanese conduct evacuation training so often that they are well prepared for a disaster. It would be one of the reasons to reduce disaster victims and damages in Japan. However, as already seen in *tsunami tendenko*, victims evacuated based on social trust which has established in families and communities over a long period of time. In this context, the existence of social capital is re-recognised in Japan, and social media could play a great role in enhancing it.

However, communication through social media in Japan experienced some problems. At first, the information exchanged via social media minimised social context cues under the severely restricted conditions. Decreasing social context cues lead to remove restrictions on communication substantively (Sproull and Kiesler 1991). Therefore, information communicated tends to be extreme and there is a risk of group polarisation. For instance, a particular social norm might accrete its influence on people, and advance self-stereotyping among them through social media.

What is worse is that unified social norms without tolerance might force people to develop a prejudicial solidarity and identity. Emphasising particular identities and norms might raise the social risk of the exclusion of others. Depending on the way of using social media, it could either help people or constrain them. In order to neutralise the negatives and strengthen the positives of social media, we need to create, manage and use social media with its various social aspects as well as ethical aspects. Also, we have responsibility to develop and keep social media in good shape because it can contribute greatly, not only to people's personal need but also

[7] http://kokunai.nihon-kankou.or.jp/, http://ganba-nippon.com/,

to social capital that is a basis for solidarity, freedom and democracy. We need to carefully observe subsequent developments and discern the impact of social media continuously.

Acknowledgement This research was supported by a grant from The MEXT Programme for Strategic Research Bases at Private Universities (2012–16) project "Organisational Information Ethics (S1291006)" and Grant-in-Aid for Young Scientists (B) "Research on Social Responsibility of Online Media Companies and Scholars in Exploring Human Behaviour (16K16169)".

References

Asai, Ryoko. 2011. "Social Media as a Tool for Change." Paper presented at ETHICOMP 2011 (*Proceedings*: 44–50), September 14–16, at Sheffield Hallam University.

Coleman, James S. 1987. "Norms as Social Capital." In *Economic Imperialism: The Economic Approach Applied Outside the Field of Economics*, edited by Gerard Radnitzky and Peter Bernholz: 133–155. New York: Paragon House Publishers.

Coleman, James S. 1990. *Foundations of Social Theory*. Cambridge, MA: Harvard University Press.

Editorial Office of Computer Technology コンピューターテクノロジー編集部. ed. 2011. *Earthquake Disaster and Nuclear-driven Damage in the IT Age* IT時代の震災と核被害. Tokyo 東京: Impress Japan インプレスジャパン.

Hayashi, Nobuyuki 林信行, and Yamaji Tatsuya 山路達也. 2012. *The Great East Japan Earthquake, Information, Internet and Google* 東日本大震災と情報,インターネット,Google. Available at http://www.google.org/crisisresponse/kiroku311/ (Accessed 18 July 2017).

Ichikawa, Hiroshi 市川浩. 1992. *The Body as Spirit* 精神としての身体. Tokyo 東京: Kōdansha 講談社.

Kobayashi, Akihito 小林啓倫. 2011. *Disaster and Social Media* 災害とソーシャルメディア. Tokyo 東京: Mainichi Communications 毎日コミュニケーションズ.

Lin, Nan. 2001. *Social Capital: A Theory of Social Structure and Action*. New York: Cambridge University Press.

Ministry of Internal Affairs and Communications. 2011. *2010 White Paper Information and Communications in Japan*. Tokyo: Gyosei.

Ministry of Internal Affairs and Communications. 2016. *2016 White Paper Information and Communications in Japan*. Tokyo: Gyosei.

Ministry of Land, Infrastructure, Transport and Tourism (2006). *White Paper on MLIT in Japan, 2006*. Tokyo: Gyosei.

Miyata, Kakuko 宮田加久子. 2005. *Media to Build Up Social Bond* きずなをつなぐメディア——ネット時代の社会関係資本. Tokyo 東京: NTT Publishing NTT出版.

Nagel, Thomas. 1970. *The Possibility of Altruism*. Princeton, NJ: Princeton University Press.

National Police Agency 警察庁. 2017. "The Damage Situation and Police Actions at the Time of the Great East Japan Earthquake in 2011 平成23 年(2011 年)東北地方太平洋沖地震の警察措置と被害状況." Available at https://www.npa.go.jp/news/other/earthquake2011/pdf/higai-jokyo.pdf (Accessed 18 July 2017).

Neubig, Graham, and Kevin Duh. 2013. How Much Is Said in a Tweet? A Multilingual, Information-theoretic Perspective. *AAAI Spring Symposium: Analyzing Microtext*: 32–39. Available online: http://www.aaai.org/ocs/index.php/SSS/SSS13/paper/view/5698 (Accessed 28 July 2017)

Oldenburg, Ray. 1999. *The Great Good Place*. Cambridge, MA: Da Capo Press.

Putnam, Robert D. 1993a. *Making Democracy Work: Civic Traditions in Modern Italy*. Princeton, NJ: Princeton University Press.

Putnam, Robert D. 1993b. "The Prosperous Community: Social Capital and Public Life." *American Prospect* 13: 35–42.

Putnam, Robert D. 1995a. "Bowling Alone: America's Declining Social Capital." *Journal of Democracy* 6.1: 65–78.

Putnam, Robert D. 1995b. "Tuning In, Tuning Out: The Strange Disappearance of Social Capital in America." *PS: Political Science and Politics* 28.4: 664–683.

Sproull, Lee, and Sara Kiesler. 1986. "Reducing Social Context Cues: Electronic Mail in Organizational Communication." *Management Science* 32.11: 1492–1512.

Sproull, Lee, and Sara Kiesler. 1991. *Connections: New Ways of Working in the Networked Organization*. Cambridge, MA: MIT Press.

Winner, Langdon. 1995. "Political Ergonomics." In *Discovering Design: Explorations in Design Studies*, edited by Richard Buchanan, and Victor Margolin: 146–172. Chicago, IL: The University of Chicago Press.

Yamori, Katsuya 矢守克也. 2012. "Revisiting the concept of 'tsunami tendenko' 「津波てんでんこ」の4つの意味." *Journal of Japan Society for Natural Disaster Science* 自然災害科学 31.1: 35–46.

Chapter 12
Reading TANAKA Shōzō as an Ethical Person After Fukushima

CHEUNG Ching-yuen

1 Introduction

On the second day after 11th March 2011, Fukushima became the focus of the world as a reactor of Daiichi Nuclear Plant exploded. Now it is known that meltdown occurred in three units of the plant. Over 7 years have passed, but the cause of meltdown or melt-through is yet to be fully explained. Some politicians and scholars naively suggested the disaster is a "punishment from heaven."[1] Meanwhile, it is generally thought that the "natural disaster" was the result of tsunami; but as the former plant manager YOSHIDA Masao 吉田昌郎 (1955–2013) notes, before all the cooling systems failed due to tsunami, the internal structure of the reactors might have been damaged during the M9-class earthquake.[2] If this is the case, Fukushima is neither a punishment from heaven nor a natural disaster, but an error of human technology that underestimated the power of nature.

How can we understand what happened in Fukushima? We should first clarify that there are "two Fukushima," namely, the Fukushima Prefecture (*Fukushima ken* 福島県; population: 1.99 Million) and the Fukushima City (*Fukushima shi* 福島市; population: 280,000). The Fukushima Daiichi Nuclear Plant is not located in Fukushima

The original version of this chapter was revised: MS Mincho font was updated throughout the book for Japanese and Chinese characters. The correction to this book is available at https://doi.org/10.1007/978-3-319-59027-1_13

[1] Examples are former Tokyo Governor Mayor Ishihara, and Buddhist scholar Sueki Fumihiko.

[2] "Reading *the Yoshida Testimony* 吉田調書を読み解く," http://www.huffingtonpost.jp/foresight/yoshida-testimony_b_6828816.html?ncid=fcbklnkjphpmg00000001. In Japanese, Accessed 1 May 2017.

CHEUNG Ching-yuen (✉)
Department of Japanese Studies, Chinese University of Hong Kong,
Shatin, Hong Kong, China

Thomas Taro LENNERFORS, MURATA Kiyoshi (eds.), *Tetsugaku Companion to Japanese Ethics and Technology*, Tetsugaku Companions to Japanese Philosophy 1,
https://doi.org/10.1007/978-3-319-59027-1_12

City; rather, it is located along the coastline of Fukushima Prefecture, known as Hamadōri 浜通り.[3] Hamadōri is an area with beautiful nature. In 2013, I visited Fukushima Prefecture with some teaching staff and students from Chinese University of Hong Kong. Saulius Geniusas from the department of philosophy notes,

> Here we are closed to the power plant. I am probably the first Lithuanian to come here, and you would think we came to see this gorgeous nature around. But this is a heavily polluted place. So the contrast is remarkable… I keep on thinking about the contrast between the beauty of the place and the severity of the disaster. I know many people who would come here and say this is at least one of the most beautiful places they have visited. The roads are remarkable. Here you can see the Pacific Ocean. The villages we passed are old and seem to be well established. And even now you can see so few people, so everything is so clean and in order. This contrast is very hard to be comprehended. How a place can be so beautiful and could be so unsafe and unlivable? The only way for the people to live here is self-forgiveness and the denial of what everything is happening. You have to somehow deceive yourself, "everything is okay; nothing is really taking place." Because if you do not do that, you will not be here.[4]

It could be rather difficult to imagine a place on this planet where it is the most beautiful but at the same time the most polluted. This lack of imagination, or more precisely the incomprehensibility of what happened in Fukushima, can be explained in the following historical context: The apparently beautiful Japanese environment is indeed a victim of Japan's project of modernization. Therefore, it is important to re-examine the history of pollution in Modern Japan, and see how people reacted to major environmental disasters.

In this article, I will discuss the case of Tanaka Shōzō 田中正造 (1841–1913). Although his name is not found in *Japanese Philosophy: A Sourcebook*, Tanaka is regarded as a social activist and the pioneer of democratic movement in Japan. Here, I shall read Tanaka as an ethical person who develops an ethics to care about nature as well as human beings.

2 Tanaka Shōzō and the Ashio Copper Mine Pollution

My reading will begin with the infamous Ashio Copper Mine (*Ashio dōzan* 足尾銅山) pollution case, which is regarded as one of the most disastrous cases of environmental destruction in modern Japan. Ui Jun 宇井純 rightly notes, "In 1868, the newly established Meiji government of Japan made the modernization of the country by increasing military strength and expanding industrial production its first national priority" (Ui 1992). The very ideas of Japan's modernization are "rich

[3] Hamadōri is one of three areas of Fukushima Prefecture. The other two areas are Aizu 会津 and Nakadōri 中通り.

[4] The Chinese Translation of the message is published in *Kitetsu: A Philosophical Magazine for Everyone* (2014), supported by knowledge transfer fund of Chinese University of Hong Kong. Details of the project can be found here: http://www.orkts.cuhk.edu.hk/knowledge-transfer-initiatives/project-highlights/228-restoration-after-3-11-pain-suffering-memory-and-trauma-kpf13icf18. Accessed 1 May 2017.

nation, strong army" (*fukoku kyōhei* 富国強兵) and "increase production, encourage industry" (*shokusan kōgyō* 殖産興業). In order to accumulate wealth for the state, copper became an important means in trading with foreign countries. The Ashio Copper Mine, discovered in 1608, is located in Tochigi Prefecture (*Tochigi ken* 栃木県; Southwest of Fukushima Prefecture) (Strong 1995: 65). It has been producing copper for the Tokugawa shogunate, but after Meiji Restoration in 1868, it was owned by an entrepreneur called FURUKAWA Ichibei 古川市兵衛 (1832–1903). From 1877 to 1891, the production of copper increased rapidly from 77 to 7486 tons per year (Strong 1995: 66).

Nonetheless, it is just the beginning of a large-scale environmental disaster. During the production of copper, sulfuric acid is released as a by-product. Sulfuric acid in the form of acidic gas killed the trees and plantations near the copper mine. In 1884, Ashio became the mine with the highest copper production in Japan, but in the same year, "the entirety of the once-forested areas around the Ashio refinery had been biologically destroyed" (Ui 1992). In 1885, it is recorded that "[t]he use of a rock-crushing machine and a steam-operated pump in the Ani mine greatly increased production but led to massive fish kills in the Watarase River."[5] It is the effect of copper mine pollution, which contaminated the river system from the upper stream to the lower stream.

Born in Tochigi Prefecture, TANAKA Shōzō developed his career as a politician and managed to be elected in Tochigi Prefectural Assembly as well as the House of Representatives. He has been well aware of the pollution issues in copper mine. As early as in September 1891, Tanaka uses the term "copper poisoning" in his letter and diary. Later during the Second Diet Meeting in December of the same year, he sent an inquiry to the government on the Ashio Copper Mine issue. He argued that the Ashio Copper Mine was causing severe environmental problems not only to agriculture areas nearby, but also to bamboo trees along the riverbanks. He asked the government how to help people who are suffering and to prevent future damages.[6] The respondent from the government was MUTSU Munemitsu 陸奥宗光 (1844–1897), Minister of Agriculture and Business at that time. Mutsu tried to avoid the responsibility of the government by suggesting that the reason of pollution is unknown, and hence no conclusion could be drawn.[7]

In 1892, Tanaka filed another inquiry to the government during the Third Diet Meeting. He quoted Mutsu that the government ordered the mining company to prevent the leak of minerals, thus affirming that the minerals from the copper mine were poisonous.[8] He also quoted a report by the Agricultural University (currently Faculty of Agriculture of the University of Tokyo), in which it was concluded that the Ashio Copper Mine is the source of water pollution in the area.[9] Since the causal

---

[5] Ibid.

[6] 「足尾銅山鉱毒の儀につき質問書」, Tanaka (2004), Vol. 1, p. 99 (Hereafter abbreviated as TSB 1:99).

[7] 「ソノ被害ノ原因ニ就テハ未ダ確実ナル試験ノ成蹟ニ基ケル定論ノアルニアラズ」, TSB 1: 106.

[8] 「足尾銅山鉱毒の儀につき質問書」, TSB 1:106.

[9] 「足尾銅山工業所排出水ノ渡良瀬川ニ入ルモノ有毒物ヲ含有スルコトマタ事実ナリ云々」, TSB 1:107.

relationship between the copper mine and the pollution was confirmed, the government could no longer avoid the responsibility. KŌNO Togama 河野敏鎌 (1844–1895), Minister of Agriculture and Business at that time, answered that the government admitted the Ashio Copper Mine was the cause of damage to nearby farmland, but that it was not affecting the general public. Therefore, it was not necessary for the company to terminate the mining operation.[10]

The situation became worse in September 1896, when a torrential rain caused a massive flood in the areas. In his diary, Tanaka reported that riverbanks were overflowed and there were many victims.[11] It was clear that the lack of planation results in flooding, but the worst scenario became reality as the flooded area was heavily contaminated by the Ashio mine pollutants. It is estimated that "[t]he loss sustained was about 23 million yen, which was eight times the annual income of the Ashio copper mine" (Ui 1992). In order to help the people there, Tanaka set up an office in Unryū Temple, which is a Sōtō Buddhist Temple located in the center of the polluted area. Later, the center became the office for the campaign of closing the Ashio Copper Mine.[12] From the correspondence of Tanaka and Monk KUROSAKI Zen'ō 黒崎禅翁 of Unryū Temple, we can see Tanaka tried to gain local people's trust, and provided education for the younger generation. Tanaka writes, "without new education, there is no new thought."[13]

Many inhabitants used to be silent, but now they turned their thoughts into action. As a result, there were many demonstrations against the mining company and the government. In 1897, for example, thousands of farmers planned to march from Tochigi to Tokyo, and a few hundreds of them even managed to reach the capital. This social movement was reported in national newspapers, drawing attention of the public to the Ashio Copper Mine issue. However, a tragic event happened in 1900. Again, thousands of villagers were on their way marching from Unryū Temple to Tokyo, but at Kawamata 川俣 of Gunma Prefecture (*Gunma ken* 群馬県) they were violently confronted by the police. Many demonstrators were injured and arrested. It is known as the Kawamata Incident. As Strong writes, "No one, after all, was killed. Nevertheless the Japanese Press, to its credit, was almost unanimous in its condemnation of the government and police for their violent suppression of the demonstration, which they claimed would never have happened if the government had not failed its duty."[14] Later, an angry Tanaka wrote an inquiry to the government:

> To kill the people is to kill the nation. To despise the law is to despise the nation. There are the ways to destroy a nation. If its resources are abused, its people killed, and its laws overturned, no country can survive. What will the government do about it?[15]

---

[10] TSB 1: 108.

[11] TSB 1:141.

[12] TSB 1: 143.

[13] TSB 1: 161.

[14] Strong 1995: 116.

[15] 「民を殺すは國家を殺すなり。法を蔑にするは國家を蔑にするなり。皆自ら國を毀つなり。財用を濫り民を殺し法を亂して而して亡びざる國なし。之を奈何。右質問に及候也。」, TSB 1: 223. Translated from Strong (1995: 119) (modified).

Ironically, Tanaka received a written "reply" from the Prime Minister YAMAGATA Aritomo 山縣有朋 (1838–1922): "The point of the question [raised by Tanaka] is not well understood. No reply, therefore, is made." (Strong 1995: 121).

In 1901, Tanaka surprised the whole nation by making a "direct petition" (*jikiso* 直訴) to the Meiji Emperor. Ui reports the details of the event as follows:

> On the morning of 10 December 1901, when, after presiding at the opening of the sixteenth National Diet Upper House session, Emperor Meiji was going to his carriage, Tanaka came up to him, a written appeal in hand, shouting to him. By this action Tanaka had planned to bring the scandal of the mine poisonings into public view, hoping that one of the imperial guards would either kill or injure him. But in fact, the sergeant-at-arms fell from his horse, which had reared up in surprise, and Tanaka also stumbled and fell on his face, so he was neither killed nor injured. He was arrested on the spot and taken into custody by the police… Tanaka was released at 7.30 the same evening. Tanaka's appeal did not work as planned, but it astounded the public at large. (Ui 1992)

The "direct petition" was, however, regarded as a disrespectful act towards the emperor. Here, Strong argues it is related to Japanese culture:

> The 'direct appeal' marked the beginning of Shōzō's shift from 'madness' to 'foolishness,' from 'agitation' in the established political world on behalf of the powerless and the have-nots to the much more demanding 'folly' of an attempt to identify his life totally with theirs. The change links him to a Japanese tradition represented by such figures as Shinran, the great saint and self-styled 'bald-pate fool' of the medieval period or Ryokan, the 'demon-cratic' priest, poet and philosopher of late eighteenth and early nineteenth centuries. (Strong 1995: 135)[16]

As a long-term solution to the Ashio Copper Mine pollution, the government suggested to construct a poisoned water catchment basin in Yanaka Village (*Yanaka mura* 谷中村). At that time, Japan is on her way to military expansion. Questioning the decision, Tanaka writes, "[t]he problem of Yanaka Village is more important than the war between Japan and Russia."[17] He even suggests that "the number of people killed by violent government is more than those died in war."[18] As an anti-war activist, Tanaka writes an article titled "Against War" (*musen ron* 無戦論) in 1903. However, Japan began the war against Russia the next year. In order to work with locals, he decided to take up residence in Yanaka Village. However, Tanaka could not change the fate of Yanaka. In 1904, it is officially announced that Yanaka is no longer a village (*mura* 村). Villagers were forced to move out, and the last village houses were destroyed. Losing their homeland, many Yanaka villagers moved

---

[16] It should be noted that in 2013, Upper House lawmaker YAMAMOTO Tarō 山本太郎, who was formerly an actor but became a politician after 3.11, approached Emperor Akihito during a party hosted by the Imperial Couple in Tokyo. Yamamoto's goal of no nuke did not come true, as Sendai nuclear plant in Kagoshima Prefecture restarted in 2015, despite potential threads from volcano eruption of Mt. Sakurajima as well as Kumamoto Earthquake in 2016.

"Anti-nuclear lawmaker tries to get Emperor's attention," http://www.japantimes.co.jp/news/2013/10/31/national/politics-diplomacy/anti-nuclear-lawmaker-tries-to-get-emperors-attention/#.Vd1X-fmqpBc. Accessed 1 May 2017.

[17] 「谷中問題ハ日露問題より大問題なり」, TSB 2:19.

[18] 「戦争ニ死するものよりハ寧ろ内地ニ虐政ニ死するもの多からん」, TSB 2: 34.

to Hokkaido 北海道 with the hope of a better life, but they had to face the cold weather with little support. The formerly Yanaka is now re-named as Watarase retarding basin (*Watarase yūsuichi* 渡良瀬遊水地).[19]

Tochigi and Fukushima Prefectures are similar in the sense that they are, like many rural areas in Japan, rich in natural resources. However, Japan's modernization depends on the sacrificing of the resources there. A famous example in Fukushima is Iwaki Coal Mine (*Iwaki tanden* 常磐炭田). Founded in 1856, the mine used to provide energy for Japan's modernization, but since its closure in 1971 many people lost their jobs.[20] As the movie Hula Girls (2006) shows, they tried to earn a living in building a Hawaiian Resort. "In 1965 the planned closing of a coal mine in Iwaki (northeastern Japan) will put 2,000 people out of work with devastating effects on the community. The mining company plans to build the Hawaiian Center to promote tourism, but the idea meets with resistance by the community's union families who boycott the effort. However, a few of the young women in Joban see the call for dancers to possibly provide a more promising future."[21] One should also note that after 3.11, there is a documentary of *Fukushima Hula Girls* (2011). "This documentary focuses on the efforts of the hula performers of the Spa Resort Hawaiians in Fukushima as they tour the country to persuade their fellow citizens to visit their resort, even as the region struggles towards recovery after the Great East Japan Earthquake."[22]

Nowadays, Japan is proud of her world-leading role in the development of advanced environmental technologies such as hybrid cars and solar-powered houses. Obviously, "green innovation" is becoming the most important keyword to Japan's unfinished project of modernization.[23] However, it is shown that our nature is so vulnerable and human technology can be extremely destructive. Besides, modernization is not only on the level of technology; it is also related to social reformation and enlightenment of citizens. Therefore, environmental pollution is not just a "physical-chemical" issue; it is related to philosophy, ethics, social justice theory, etc.

Ideally speaking, environmental problems demand participation and action from the people. People should be able to change their fate. However, TANAKA Shōzō gives us a lesson: People who sacrificed their lives for the state are the sufferers. As TAKAHASHI Tetsuya 高橋哲哉 argues, there is a "sacrificial system" in which the vulnerable are forced to sacrifice for the power (Takahashi 2012). While in Okinawa people suffer with many US military bases, in Fukushima people lost their homeland due to the nuclear disaster. Takahashi mentions a proposal by Danish General Frits Holm in 1928 that in the event of an accident, "the following should constitute

[19] There is a Japanese drama about the last days of Ashio and Yanaka. Details can be found here: http://www.nhk.or.jp/dodra/ashio/index.html. Accessed 1 May 2017.

[20] http://tankouisan.jp/data/year_history/index.html. Accessed 1 May 2017.

[21] http://www.imdb.com/title/tt0768116/plotsummary. Accessed 1 May 2017.

[22] http://www.us.emb-japan.go.jp/jicc/events/fukushima-hula-girls.html. Accessed 1 May 2017.

[23] On the rhetoric of Japan's "green technology," see Jonathan Taylor (1999), "Japan's global environmentalism: rhetoric and reality," *Political Geography* 18: 535–562.

a 'suicide squad' and be sent into the reactor: the prime minister, the cabinet, the vice-minister and top officials of the Ministry of Energy, Technology, and Industry, the CEO of the power company and top officials, scientists and technical experts who promoted nuclear power. Nor can we overlook the responsibility of urban dwellers, myself included, who imposed the construction of nuclear power plants on rural areas and received the electricity thus generated."[24]

Japan did not change after the Fukushima disaster, for the sacrificial system remains the same. It is reported that workers in the nuclear plant continue to receive high dosage of radiation in the decommissioning plan of Fukushima Daiichi Nuclear Power, while there are increasingly more cases for children suffer from thyroid cancer.[25] Manu beautiful sites in Fukushima as well as nearby prefectures are being sacrificed as temporary disposal sites for contaminated soil and water. Besides, Japan has yet to confirm the location of the disposal site for nuclear wastes. In 2007, the central government called for applications from local governments about a final disposal site. There was one application from Tōyō Town (*Tōyō cho* 東洋町) in Kōchi Prefecture (*Kōchi ken* 高知県). The mayor, TASHIMA Yasuoki 田嶋裕起, faced resistance from people living in the town and he lost a subsequent election. Since Japan is situated on the Pacific seismic belt, a safe site for nuclear plant or disposal site does not exist.

3 Concluding Remarks

In his presentation to the International Olympic Committee (IOC) in 2013, Prime Minister ABE Shinzō 安倍晋三 was desperate to emphasize that Fukushima is "under control," He continued, "Some may have concerns about Fukushima. Let me assure you, the situation is under control. It has never done and will never do any damage to Tokyo."[26] He also claimed: "I explained about the water contamination in Fukushima and explained that the contaminated water was blocked." For Fukushima, the road to recovery will be time and money consuming, but people in Japan would care more about the Olympic Stadium, which was initially priced at ¥265 billion (recently cut to ¥155 billion). It is nothing but an international scandal, but no one is responsible for the high quotation.

Unfortunately, Fukushima is no longer the priority of the government and the people, as Tokyo will host the Olympic Games in 2020. How can we *understand* Fukushima? Or perhaps we may ask, how can we *remember* Fukushima? Japanese critic AZUMA Hiroki 東浩紀 and his colleagues studied the case of Dark Tourism in Chernobyl, and proposed a turn the Fukushima nuclear power plant into a sightsee-

[24] https://ceas.uchicago.edu/news/what-march-11-means-me. Accessed 1 May 2017.

[25] http://www.japantimes.co.jp/news/2017/03/31/national/boys-thyroid-cancer-casts-doubt-fukushimas-denials/#.WQqc0_mGPIU. Accessed 1 May 2017.

[26] http://japan.kantei.go.jp/96_abe/statement/201309/07ioc_presentation_e.html. Accessed 1 May 2017.

ing spot. His concern is "To make sure the memories of the disaster don't fade away." Azuma is well aware of the lack of support of the idea from local people, but he argues people living in Tokyo should be involved. He writes,

> I certainly don't butt in when it comes to other natural disasters, but the Fukushima No. 1 nuclear power plant belongs to Tokyo Electric Power Co., so I, and all the other residents of Tokyo, are without a doubt 'concerned parties'... This accident doesn't just affect Fukushima; it affects all of humanity. Even if, hypothetically speaking, the people of Fukushima were to say they don't want to see it anymore and demand the site be turned into vacant land, it would still be unacceptable.[27]

According to Azuma's plan of Tourizing Fukushima, "Dark tourism is a concept that appeared in the 1990s. It refers to a new style of tourism where individuals visit locations where tragedies such as wars and disasters occurred, such as Hiroshima広島 and Auschwitz. As a country regularly arranges school trips to Hiroshima and Okinawa 沖縄, Japan is, in fact, a leader in the field of dark tourism. One of the underlying supports to our plans is this concept of dark tourism."[28] In fact, many Japanese students make school trips to Hiroshima and Nagasaki 長崎 to remember the victims of nuclear bombs. They take photographs of remains, debris, temples and shrines, as most of the tourists would do. However, these tours are not for tourist purpose, but for an educational one. I suggest that these visits may be called "pilgrimage" (*junrei* 巡礼). Pilgrimage is a word from pilgrim or peregrine. In Latin, *peregrinus* means "foreign," and *peregre* means "abroad." It could be understood as an experience through (*per-*) a field (*ager*). Although pilgrims are not tourists, they go to a foreign but yet unhomely place to know the world and face the other (the world and the other are not merely philosophical concepts!). More importantly, it is a practice to share historical pain and preserving memories by sympathy or commiseration.[29]

I do not oppose the idea of dark tourism, but for Fukushima it may take a long period of time. Before any "sightseeing spot" is established, we *should* go for a "pilgrimage" to Fukushima. Besides the wrecked nuclear plant, there are many places worth a visit. For example, Futaba Town (*Futaba machi* 双葉町; screened in the movie *Ieji,* 2014) is a place near the Fukushima Nuclear Plants. We should listen to those who take the risk to return to the polluted land. We should also stop by Tochigi Prefecture, the homeland of TANAKA Shōzō who believes "true civilization is not to despoil mountains, not to ruin rivers, not to destroy villages, and not to kill people."[30]

In an article titled "Fukushima's history of struggle" (19 March 2014), a victim of Fukushima's nuclear disaster writes, "Our present anti-nuclear movement is similar to the one for democracy that our ancestors waged during the Meiji Era. While their dream for democracy didn't come true, their will has been handed down to us."

---

[27] *Asahi Shimbun Weekly AERA*, 7 October 2013.

[28] http://fukuichikankoproject.jp/project_en.html. Access 1 May 2017.

[29] I borrow the idea of pilgrimage from Watsuji Tetsurō's book Watsuji Tetsurō (1889–1960): *Pilgrimages to the Ancient Temples in Nara*, Honolulu: University of Hawai'i Press, 2012.

[30] http://www.cnet-ga.ne.jp/syozou/. Accessed 1 May 2017.

It is true that Fukushima is a complex issue, but it can be understood in the following way: the nation is sacrificing her people in the name of the project of modernization.[31] The author also ends her article by quoting TANAKA Shōzō:

> The politician Shozo Tanaka (1841–1913), a pioneer of Japan's ecology movement who struggled against the environmental disaster known as the Ashio Copper Mine Incident in the late 19th and early 20th centuries, wrote: "True civilization neither ruins mountains, pollutes rivers, destroys villages, nor kills people."[32]

References

Cheung, Ching-yuen 張政遠. ed. 2014. *Kitetsu: A Philosophical Magazine for Everyone* 希哲雜誌. Hong Kong 香港: Chinese University of Hong Kong 香港中文大學.

Ui, Jun, ed. 1992. *Industrial pollution in Japan*. Tokyo: The United Nations University. http://archive.unu.edu/unupress/unupbooks/uu35ie/uu35ie00.htm. Accessed 1 May 2017.

Strong, Kenneth. 1995. *Ox Against the Storm: A Biography of Tanaka Shozo*. Sandgate, Folkstone, Kent: Japan Library.

Takahashi, Tetsuya 高橋哲哉. 2012. *The System of Sacrificing: Fukushima and Okinawa* 犠牲のシステム 福島・沖縄. Tokyo 東京: Shūeisha 集英社.

Tanaka, Shōzō 田中正造. 2004. *Tanaka Shōzō Bunshū* 田中正造文集, 2 vols. Tokyo 東京: Iwanami Shoten 岩波書店. (abbreviated as TSB)

Taylor, Jonathon. 1999. "Japan's global environmentalism: rhetoric and reality." *Political Geography* 18: 535–562.

Watsuji, Tetsurō. 2012. *Pilgrimages to the Ancient Temples in Nara*. Honolulu: University of Hawai'i Press.

[31] For example, the author of the article reports a tragic story: "As for Fukushima's nuclear disaster, another resident was feeding 350 cows for food markets. After March 2011, the government ordered people like him to kill their cows. He refused. The area has been closed off, but he has been feeding more than 300 cows under economic and physical hardships for the purpose of having the cows tested for radiation effects at a medical facility in the future."

[32] http://www.japantimes.co.jp/opinion/2014/03/19/reader-mail/fukushimas-history-of-struggle/#.V0JwW_l97lU. Accessed 1 May 2017.

Correction to: Tetsugaku Companion to Japanese Ethics and Technology

Thomas Taro LENNERFORS and MURATA Kiyoshi

Correction to:
Thomas Taro LENNERFORS, MURATA Kiyoshi (eds.),
Tetsugaku Companion to Japanese Ethics and Technology**,**
Tetsugaku Companions to Japanese Philosophy 1,
https://doi.org/10.1007/978-3-319-59027-1

This book was inadvertently published with the incorrect font (Adobe Myungjo Std) used to render Japanese and Chinese characters. This font has now been changed to MS Mincho throughout the book except in Chapter 6 where original Chinese characters are changed to SimSun font.

The updated version of the book can be found at
https://doi.org/10.1007/978-3-319-59027-1

Thomas Taro LENNERFORS, MURATA Kiyoshi (eds.), *Tetsugaku Companion to Japanese Ethics and Technology*, Tetsugaku Companions to Japanese Philosophy 1,
https://doi.org/10.1007/978-3-319-59027-1_13

MIX
Papier aus verantwortungsvollen Quellen
Paper from responsible sources
FSC® C105338

If you have any concerns about our products,
you can contact us on
ProductSafety@springernature.com

In case Publisher is established outside the EU,
the EU authorized representative is:
Springer Nature Customer Service Center GmbH
Europaplatz 3, 69115 Heidelberg, Germany

Printed by Libri Plureos GmbH
in Hamburg, Germany